Integrated Crop Management Vol.12-2010

GREEN MANURE/COVER CROPS AND CROP ROTATION IN CONSERVATION AGRICULTURE ON SMALL FARMS

Miguel Angel Florentín
Marcos Peñalva
Ademir Calegari
Rolf Derpsch

Translated by
Melissa J. McDonald

PLANT PRODUCTION AND PROTECTION DIVISION
FOOD AND AGRICULTURE ORGANIZATION OF THE UNITED NATIONS
Rome, 2011

The designations employed and the presentation of material in this information
product do not imply the expression of any opinion whatsoever on the part
of the Food and Agriculture Organization of the United Nations (FAO) concerning the
legal or development status of any country, territory, city or area or of its authorities,
or concerning the delimitation of its frontiers or boundaries. The mention of specific
companies or products of manufacturers, whether or not these have been patented, does
not imply that these have been endorsed or recommended by FAO in preference to
others of a similar nature that are not mentioned.

The views expressed in this information product are those of the author(s) and
do not necessarily reflect the views of FAO.

ISBN 978-92-5-106856-4

All rights reserved. FAO encourages reproduction and dissemination of material in this
information product. Non-commercial uses will be authorized free of charge, upon
request. Reproduction for resale or other commercial purposes, including educational
purposes, may incur fees. Applications for permission to reproduce or disseminate FAO
copyright materials, and all other queries concerning rights and licences, should be
addressed by e-mail to copyright@fao.org or to the Chief, Publishing Policy and
Support Branch, Office of Knowledge Exchange, Research and Extension, FAO,
Viale delle Terme di Caracalla, 00153 Rome, Italy.

© FAO 2011

PROLOGUE

This publication is the result of activities developed within the framework of the "Soil Conservation Project MAG – GTZ", implemented from 1993 to 2001. The executing institutions of the Soil Conservation Project were the Ministry of Agriculture and Livestock (MAG) of the Republic of Paraguay and the German Technical Cooperation (GTZ).

The Spanish original has been published in Paraguay by the Ministry of Agriculture and Livestock (MAG) of the Republic of Paraguay, the Directions of Agrarian Research and Extension (DIA/DEAG) and the German Technical Cooperation (GTZ). The present English version has been translated by Melissa J. McDonald under the review of the original authors and is published in the FAO-AGP integrated crop management series.

The information utilized in this work is based on research done principally at the Choré Experimental Station of the Direction of Agricultural Research (DIA/MAG). Also used are experiences developed in pilot areas, especially in Paraguarí, Edelira, Minga Guazú, Caaguazú, Guairá, Caazapá, and San Pedro, among others. Diffusion and extension activities were carried out through the Direction of Agrarian Extension (DEAG), cooperatives, farmers' associations, self-help groups, etc.

The objective of this publication is to offer a reference material for extensionists, professors, agronomy students, technicians in general, and for farmers themselves. Through information that is up-to-date and richly illustrated, it strives to facilitate the adoption and diffusion of No-Tillage, the use of green manures, and the practice of crop rotation on small farms.

The wealth of this work is that it brings together the experiences of farmers, extensionists, and researchers in a way that is simple, understandable, and practical. It describes the principal species of green manures and, at the same time, informs in detail how to insert green manures into small farm production systems according to soil fertility and major crops. It also deals with the residual effect of green manures on main crops and analyzes the economic implications of these practices. Furthermore, it describes the results obtained in the recuperation of extremely degraded soils. Finally, this work strives to show the way to achieve an agriculture that is more productive, profitable, competitive, and sustainable, with the objective of improving the quality of life of rural families.

Rolf Derpsch	**Ken Moriya**
Principal Advisor, GTZ	National Counterpart

Soil Conservation Project MAG – GTZ
San Lorenzo, Paraguay, 2001

In this publication, the terms "green manure" and "cover crop" are used as synonyms. In Conservation Agriculture, the residues of green manure/cover crops (GMCCs) are always left on the soil surface, and are incorporated biologically rather than with tillage implements.

Conservation Agriculture, with the use of green manure/cover crops and crop rotation, is the way to achieve agricultural sustainability on small farms.

CONTENTS

ii   Prologue

CHAPTER 1
1   Introduction

CHAPTER 2
3   Characterizacion of agricultural systems on small farms in paraguay according to climate and soils

CHAPTER 3
9   Green manure/cover crops (GMCCS)
9   Objectives and benefits of green manure/cover crops
10   Importance of green manure/cover crops in small farm systems
10   *Soil cover*
13   *Maintenance and/or accumulation of organic matter*
14   *Addition of nitrogen (biological fixation) and nutrient recycling*
15   *Weed control*
17   *Increase and maintenance of crop productivity and decrease in production costs*
18   General qualifications that should be met by green manure/cover crops
19   Species of green manure/cover crops most adapted to small farming systems
19   *Spring-Summer Species*
34   *Fall-winter species*
45   *Summary of characteristics and recommendations for use of green manure/cover crops on small farms*
45   Association (mixtures) of green manure/cover crops
55   Ways to include green manure/cover crops in small farm production systems
55   *Green manure/cover crops (GMCCs) associated with annual crops*
57   *Green manure/cover crops (GMCCs) in succession with annual crops*
58   *Green manure/cover crops (GMCCs) in association with perennial crops*
59   *Use of perennial and semi-perennial green manure/cover crops (GMCCs) to recuperate degraded soils*

60	Alternative uses for green manure/cover crops
60	*Animal and human nutrition*
62	*In beekeeping*
63	*Utilization for firewood*
63	Residual effects of green manure/cover crops on main crops
64	*Residual fertilizer effect of green manures on cotton*
66	*Residual fertilizer effect of green manures on corn*
68	*Residual fertilizer effect of green manures on tobacco*
68	*Residual fertilizer effect of green manures on cassava*

CHAPTER 4
71	**Crop rotation**
71	General considerations
72	Advantages of crop rotation
73	Aspects to take into account in order to establish crop rotations
74	Proposed production systems with crop rotations
74	*Crop rotation in Conservation Agriculture for soils of moderate fertility*
78	*Crop rotation in Conservation Agriculture to recuperate extremely degraded soils*
84	*Other proposals*

CHAPTER 5
87	**Advances and obstacles in the adoption of Conservation Agriculture on small farms**
93	**Annex**
95	**References**

ACKNOWLEDGEMENTS

The authors would like to thank the Ministry of Agriculture and Livestock (MAG in Spanish) and its Directions of Agrarian Extension (DEAG) and Agricultural Research (DIA), as well as the German Technical Cooperation (GTZ), institutions that made this publication possible.

We would especially like to thank the farmers who collaborated with their work on farms where experiments and demonstration plots were installed, and that served as the basis for some of the proposals raised.

We would also like to thank the technicians and field personnel of the "Soil Conservation Project MAG-GTZ", and employees of the DEAG, principally of the Paraguarí and Ybycuí regional offices, and of the DIA of the Choré Experimental Station.

We are especially grateful to agronomists Juana Caballero and María Noce for their revision of the publication, which without a doubt contributed to the improvement of its quality, and to agronomist Alba Esteche for her appraisals after having read it.

To technical extensionist Néstor Paniagua, for his/her valuable field data that served to enrich the publication.

To technician Magin Meza, for his/her excellent contribution of experiences and knowledge, which helped to enrich and improve the technical quality of the work.

CHAPTER 1
Introduction

Soil degradation on small farms in the Eastern Region of Paraguay is the principal cause of a continuous decrease in crop production. The consequences of this are reduced economic income and increased poverty among rural families.

One of the principal reasons for this fact is the continuous utilization of inadequate methods of soil management, including the burning of vegetative residues, excessive tillage, and monoculture. The exposure of bare soil to climatic agents (high temperatures, torrential rains) accelerates the soil degradation process, as they cause excessively rapid decomposition of biomass and favor the erosion and leaching of nutrients.

The decrease in productivity is closely tied to a decline in the levels of soil organic matter. In poor soils, it is organic matter that determines the improvement of physical aspects, water retention, and biological activity, as well as the storage and slow release of nutrients. These aspects are most significant in sandy soils. Therefore, in order to maintain soil productivity in agricultural systems on small farms (where chemical fertilizers are normally not used) biomass is shown to be an essential element, due to the fact that it permits nutrient recycling and controls the microbial population that maintain favorable soil properties. One great difficulty in relation to the maintenance of biomass in tropical and subtropical climatic regions is that it breaks down much more rapidly than the capacity of conventional agricultural systems to replace it.

The strategies used by technicians and farmers to counteract the reduced yields caused by soil degradation are often based on the use of large quantities of inputs (fertilizers and pesticides), which results in high production costs. The apparent absence of agricultural systems for sustained productivity within the scope of small farmers is often due, not to lack of technology or to the low yield potential of traditional varieties. Rather, this situation is due to limited knowledge, or lack of awareness, on the part of technicians and farmers about practices that function in harmony with the environment in tropical and subtropical regions. At times, the absence of operative conditions (availability of credit, seeds, machinery, etc.) is a limiting factor in the development of this process.

Management measures to maintain soil fertility on small farms in Paraguay should be oriented toward the utilization of practices that maximize biomass production while minimizing its decomposition. In this sense, crop rotation, together with the use of green manure/cover crops and No-Tillage, form part of a technological strategy that has been proven by research and farmer practice to be efficient and economically viable, and that has as its objective the increase and conservation of soil organic matter. These practices together are referred to as "Conservation Agriculture".

> Crop rotation and green manure/ cover crops constitute a technology that is appropriate and essential to achieve sustainable agricultural production.

CHAPTER 2
Characterizacion of agricultural systems on small farms in Paraguay according to climate and soils

Knowledge of local agroecological and socioeconomic characteristics is essential for the planning of strategies for the use and management of soils in a region. The diagnosis of production systems consists of gathering information on production factors and components, considering such indicators as: climate, degree of soil degradation, actual land use, available infrastructure, the conservation practices utilized, integration of diverse production activities, socioeconomic characteristics, etc.

2.1 CLIMATE

The predominant climate in the Eastern Region of Paraguay is subtropical, humid, mesothermic, with no dry season, and classified as Cfa according to Köppen. Spring and summer are the seasons when torrential rains (over 60 mm/h) and elevated temperatures of around 41 °C frequently occur. In autumn-winter, precipitation and temperatures are lower than during the rest of the year, and with the risk of sporadic frost (Figure 1). Minimum temperatures may occasionally be lower than 0 °C.

Meteorological data for the principal regions of Eastern Paraguay are presented in Table 1. Average annual precipitation varies from 1,551 to 1,707 mm. Average annual temperatures fluctuate around 21.1 to 23.0 °C.

Rainfall distribution and monthly temperature values are similar for the principal regions mentioned. Average monthly precipitation ranges between 50 and 196 mm, July being the month with the lowest rainfall of the year. Average monthly temperatures vary between 15.5 and 27.1 °C, June being the coldest month and January the hottest.

FIGURE 1
In Paraguay there are occasional frosts that affect some green manure/cover crops that are more sensitive, such as the grey-seeded mucuna (foreground), while others such as lablab (background) are less affected.

TABLE 1
Average monthly precipitation (mm) and temperature (°C) for the principal regions of Eastern Paraguay.

Region	Month												Av.
	J	F	M	A	M	J	J	A	S	O	N	D	
North[1]													
Precipitation	132	167	125	160	141	103	50	68	112	168	161	165	1.551
Av. Temp.	27.1	26.6	25.9	23.3	20.2	18.2	18.2	20.1	21.2	23.9	25.1	26.6	23.0
Central[2]													
Precipitation	186	154	144	180	128	96	60	82	104	162	196	177	1.670
Av. Temp.	26.5	26.1	25.1	22.3	19.5	17.5	17.6	18.8	20.5	22.7	24.4	25.8	22.2
East[3]													
Precipitation	148	130	117	141	138	133	73	102	125	174	136	171	1.590
Av. Temp.	26.1	25.6	24.7	21.6	18.4	16.5	17.8	17.9	19.2	22.1	23.8	25.4	21.6
South[4]													
Precipitation	156	150	154	157	142	140	103	122	129	161	152	143	1.707
Av. Temp.	26.3	25.5	24.0	21.3	18.3	15.5	16.0	16.9	18.9	21.2	22.9	25.9	21.1

Source
1 Choré Experimental Station, Department of San Pedro (23 year average). Latitude: 24° 10′ S., longitude: 56° 37′ W., altitude: 220 meters above sea level.
2 National Agronomic Institute (IAN) in Caacupé, Department of Cordillera (38 y 42 year averages for precipitation and temperature, respectively). Latitude: 25° 24′ S., longitude: 57° 06′ W., altitude: 228 meters above sea level.
3 Agricultural Technology Center in Paraguay (CETAPAR) in Iguazú, Department of Alto Paraná (28 year average). Latitude: 25° 27′ S., longitude: 55° 02′ W., altitude: 280 meters above sea level.
4 Regional Agricultural Research Center (CRIA) in Capitán Miranda, Department of Itapúa (30 year average). Latitude: 27° 17′ S., longitude: 55° 49′ W., altitude: 200 meters above sea level.

In the eastern and southern regions of Eastern Paraguay (Departments of Alto Paraná and Itapúa) the climate is colder than in the rest of the country. In the northern region (Department of San Pedro) rainfall is less and periods of water deficiency occur with greater frequency.

2.2 SOILS

Agricultural areas in the Eastern Region of Paraguay are located over two types of soil: sandy soil derived from sandstone and clay soil from basalt.

Eastern Paraguay's sandy soils are mostly located in the central and northern regions (along the Paraguay River). They are principally red-yellow (Figure 2) and dark red Podzols[1] (Ultisols and Alfisols[2]). The clay soils are principally Clay Alfisols (Figure 3) and Latosols (Alfisols, Ultisols and Oxisols) that are found in a band that borders the Paraná River.

In general, agricultural soils originate as forest land and are very fertile in the first years of use. This is due principally to the high content of organic matter, which allows for good crop yields. Under the traditional cropping system - which implies deforestation, burning (Figure 4), cutting weeds with

---

[1] Brazilian soil classification system.
[2] American soil classification system.

machete (Figure 5), and repeated and continuous tilling - soil fertility has decreased and, in consequence, crop production has gone down (Figures 6 and 7).

Clay soils, in spite of having a natural fertility superior to that of sandy soils, show the same tendency to degrade over the years when managed in a conventional system.

FIGURE 2
Profile of sandy soil (red-yellow podzolic), predominant in the Central and Northern region of Eastern Paraguay.

FIGURE 3
Profile of clay soil (clay alfisol) that is predominant along the margin of the Paraná River (Departments of Itapúa, Alto Paraná, Canindeyu).

> Traditional systems of tillage and the burning of plant residues impoverish the soil and the people who cultivate it.

FIGURE 4
In Paraguay there are occasional frosts that affect some green manure/cover crops that are more sensitive, such as the grey-seeded mucuna (foreground), while others such as lablab (background) are less affected.

FIGURE 5
The employment of labor to cut weeds in traditional cropping systems increases production costs.

2.3 AGRICULTURAL SYSTEMS

The majority of small farms in Paraguay have 5 to 20 ha (54%), a lesser proportion correspond to farms of 1 to 5 ha (37%) and scarcely 9% of farms have less than 1ha (National Farm Census, 1991). These proportions, with some variation, could be extended to the Eastern Region.

Agricultural systems are characterized as being not intensive; they utilize almost exclusively spring-summer crops, and are managed manually or in a way that is semi-mechanized (with animal traction). Land is generally not utilized during the fall-winter period, and weeds are left to grow freely until the start of the next agricultural season. The principal crops for home use are corn, cassava, cowpeas, peanuts, and sweet potato. The most important annual cash crops are cotton, corn, and cassava. In some regions, other crops grown commercially are sugar cane, soybeans, tobacco, yerba mate (Ilex paraguariensis), Phaseolus beans, sesame, peas, and others.

Production systems on small farms in Paraguay's Eastern Region are quite diverse and complex. However, at the present time it is possible to identify zones that are similar with respect to climate, soil texture, and fertility. These zones can be used as recommendation units, to introduce production systems with common strategies for soil use and management. Three principal zones stand out, which are:

1) Zone of moderately fertile sandy soils (7 to 10 years of use) with an average organic matter content of 1.2%, found principally in the Departments of San Pedro, Caaguazú, Concepción, and Caazapá.
2) Zone of very degraded sandy soils (over 15 years of use) with less than 1% organic matter, generally compacted (hardpan at a depth of 10 to 15 cm), found principally in the Departments of Paraguarí, Central, Cordillera, and Guairá.
3) Zone of clay soils with moderate to high fertility (organic matter from 2 to 3%), found principally in the Departments of Alto Paraná and Itapúa, where the somewhat colder and rainier climate of the Eastern Region occurs.

The definition of soil management recommendations by zones having similar characteristics presents the following advantages: it rationalizes the application of financial and human resources, reduces operational costs in the implementation of common practices, promotes sectoral and institutional integration, encourages participation in group discussion, stimulates the organization of farmers, and more. However, **in order to zonify the recommendations, it should also be taken into consideration that:**

- Transition zones exist, with soil and climatic characteristics that are intermediate to those already considered, such as in the Departments of Misiones, Ñeembucú, Canindeyú, part of Caazapá, and part of Guairá. Alternative production systems should be adapted for each situation.
- The production systems identified could occur anywhere in the Eastern Region of Paraguay, not exclusively in the regions mentioned.

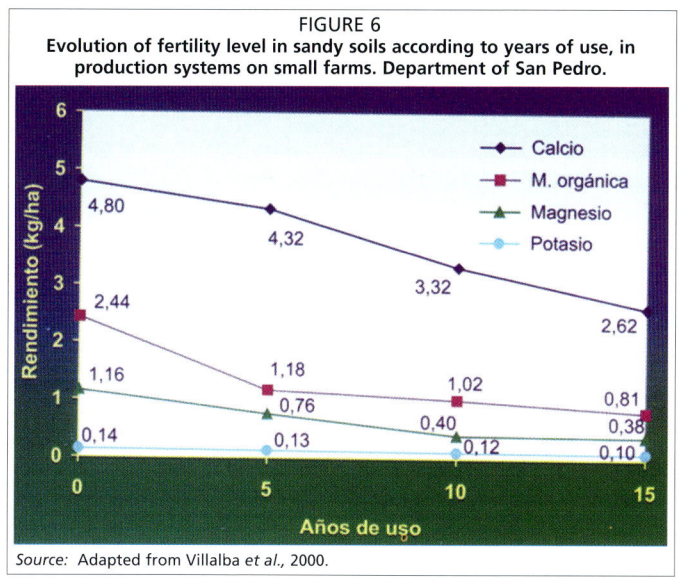

FIGURE 6
Evolution of fertility level in sandy soils according to years of use, in production systems on small farms. Department of San Pedro.

Source: Adapted from Villalba et al., 2000.

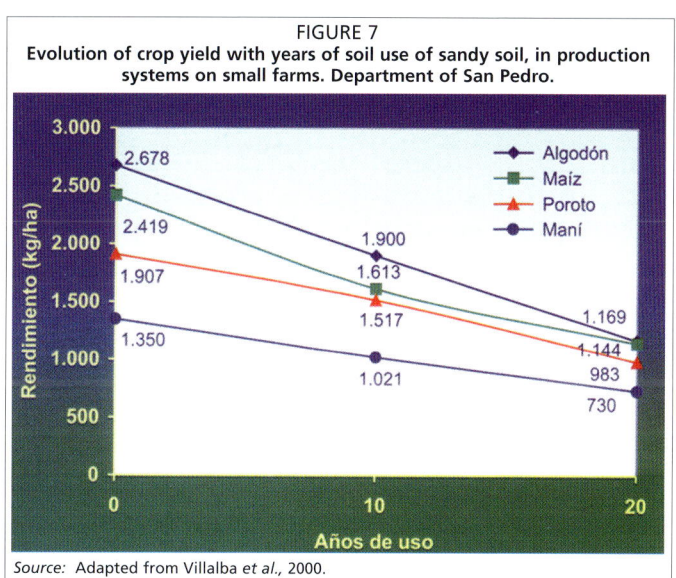

FIGURE 7
Evolution of crop yield with years of soil use of sandy soil, in production systems on small farms. Department of San Pedro.

Source: Adapted from Villalba et al., 2000.

CHAPTER 3
Green manure/cover crops (GMCCs)

In this publication, the green manures referred to are always utilized in Conservation Agriculture.

3.1 OBJECTIVES AND BENEFITS OF GREEN MANURE/COVER CROPS

Green manure/cover crops (GMCCs) are plants that are grown in order to provide soil cover and to improve the physical, chemical, and biological characteristics of soil. GMCCs may be sown independently or in association with crops.

In general, green manure/cover crops are used to pursue the following **objectives:**
- Provide soil cover for No-Tillage (reduces water evaporation and soil temperature, and increases water infiltration).
- Protect soil from erosion.
- Reduce weed infestation.
- Add biomass to soil (in order to accumulate soil organic matter, add and recycle nutrients, feed soil life).
- Improve soil structure.
- Promote biological soil preparation (Figures 8 and 9).
- Reduce pest and disease infestation.

By carrying out these functions, green manure/cover crops offer the following **benefits:**
- Increase economic return (when adequately chosen).
- Reduce need to use herbicides and pesticides.
- Increase yield and improve quality of the following crops.
- Prevent soil erosion.
- Conserve soil humidity.

FIGURE 8
Jack bean, after being flattened with a knife-roller, leaves the ground free of weeds and decompacted. The bar penetrates easily.

FIGURE 9
In fields without green manure/cover crops (in winter fallow), weeds proliferate and soil remains compacted. The bar penetrates with difficulty.

- Maintain or increase soil organic matter content.
- Provide nitrogen to the soil.
- Improve soil fertility.
- Reduce fertilization costs.

3.2 IMPORTANCE OF GREEN MANURE/COVER CROPS IN SMALL FARM SYSTEMS

3.2.1 Soil cover

Since the climate of Paraguay's Eastern Region is predominantly warm and humid, one of the principal benefits of green manure/cover crops is the improvement of soil cover (live or dead). This has favorable effects on the physical, chemical, and biological properties of soil.

FIGURE 10
Tilling with a plow leaves the soil bare, favors erosion, lowers organic matter levels, and leads to soil degradation.

In conventional systems (tillage with plow, burning), in which bare soils predominate, the direct impact of raindrops breaks soil aggregates in the surface layer, resulting in the obstruction of the soil's pores. This causes the sealing of the soil surface and impedes the infiltration of water, which then runs off the surface, carrying with it part of the soil and producing what is known as water erosion (Figure 10).

In systems that promote soil protection, the principal effect of cover is to reduce erosion, either by preventing the direct impact of raindrops on soil and/or by reducing the velocity of surface runoff (Figure 11).

FIGURE 11
Grey-seeded mucuna leaves an excellent dead cover that protects soil from erosion, leaves it free of weeds, and makes No-Tillage viable.

Cover also helps maintain soil humidity through shading and control of the thermal regime (lowers temperature, therefore reduces evaporation of water from the soil and increases water infiltration). These conditions also favor greater biological activity of the soil's flora and fauna (improves nodulation of legumes, increases the quantity of earthworms, etc.).

The temperature of bare soil in Paraguay's summer rapidly surpasses 30 °C in the early morning, and can reach nearly 60 °C (Figure 12). Soil temperatures higher than 35 °C limit to a great degree the absorption of water and nutrients by roots, and the plant's metabolism practically stops. Also, with these temperatures the activity of soil organisms (rhizobia, mycorrhizae) is significantly reduced or paralyzed. Soil temperatures should never exceed 40 °C.

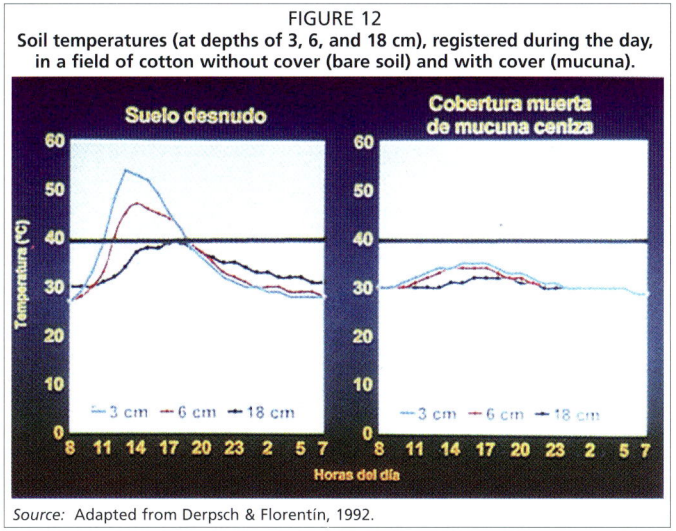

FIGURE 12
Soil temperatures (at depths of 3, 6, and 18 cm), registered during the day, in a field of cotton without cover (bare soil) and with cover (mucuna).

Source: Adapted from Derpsch & Florentín, 1992.

Plant cover, whether dead or alive, creates a microclimate that reduces the speed at which organic matter decomposes (mineralizes), favoring its accumulation in the soil. Furthermore, cover reduces the loss of organic matter from erosion. The degree of protection that a GMCC's live cover can offer the soil will depend principally on its initial speed of growth, growth habit (upright, creeping, etc.), the degree or volume of growth, as well as the duration of the plant's cycle.

Once flattened, or once the GMCC's cycle has ended, the durability of the dead cover over soil depends on the quantity and distribution of biomass produced. Likewise, it depends on the speed at which the organic matter decomposes, which, in turn, depends on the Carbon/Nitrogen ratio and on the lignin content of the residues (which varies by species). It also depends on environmental conditions that affect microbial activity (temperature, humidity, soil oxygenation, pH, etc.).

The percentages of soil cover of different species of green manure/cover crops, winter as well as summer, utilized in production systems on small farms and evaluated under the sandy soil conditions of the Choré Experimental Station, are presented in Tables 2 and 3.

TABLE 2
Percentage of soil cover during the vegetative period of principal winter green manure/cover crops, evaluated in agricultural systems for small farms (average of 3 agricultural periods). Choré Experimental Station.

Species	% of soil cover (sown in April) Days after emergence						
	15	30	45	60	75	90	105
Black oats + Hairy Vetch	15	30	35	84	93	99	100
Black oats + White Lupine	20	25	36	91	96	95	100
Black oats	17	30	35	69	84	90	100
Oilseed radish	12	26	43	73	88	95	99
Hairy Vetch	14	26	30	71	82	86	98
Bitter White Lupine	13	24	30	41	81	83	97
Sunflower	12	25	33	49	87	94	92
Triticale	12	27	28	47	69	76	77
Sweet White Lupine	10	25	29	33	60	57	48
Peas	17	27	34	53	69	67	45
Winter fallow	22	25	36	47	52	43	43

Source: Florentín, 1999 (unpublished).

TABLE 3
Percentage of soil cover during the vegetative period of several summer green manure/cover crops, evaluated in agricultural systems for small farms (average of 3 agricultural periods). Choré Experimental Station.

Species	% of soil cover (sown in October) Days after emergence													
	45	60	75	90	105	120	135	160	180	195	225	255	270	290
Calopo	15	75	100	100	100	100	90	90	90	90	90	80	80^3	80^3
Jack bean	60	85	75	65	65	65	55	50	25	25	20	15	15	10
Butterfly pea	15	40	60	82	100	100	100	100	100	100	100	90	90	90
Sunnhemp	20	30	45	45	45	40	40	35	30	20^3	-	-	-	-
Crotalaria paulina	25	40	55	65	83	80	95	90	75	70	60	45	45	40
Crotalaria striata	20	50	65	65	85	65	80	80	80	75	60	40	30	30.
Lab-lab[1]	65	95	100	98	98	95	90	90	90	90	90	80	80	80
Lab-lab[2]	75	100	100	100	99	90	95	90	90	85	50	25	25	20
Pigeon pea	35	70	80	-	95	95	90	80	55	50	50	40	40	35
Grey-seeded mucuna	70	95	95	95	100	98	95	85	85	80	80^3	80^3	80^3	80^3
Dwarf mucuna	50	95	95	89	98	95	90	90	35^3	30^3	25^3	25^3	-	-
Cowpea var. Colorado	63	85	80	45	60	40^3	-	-	-	-	-	-	-	-
Cowpea var. Tupí	60	100	100	92	97	98	90	65	15^3	-	-	-	-	-
Tephrosia	10	10	10	15	20	25	40	40	75	75	75	65	50	50

[1] White seed. [2] Black seed. [3] Cover of dry residues after the end of plant's cycle.
Source: Adapted from Derpsch & Florentín, 1992.

3.2.2 Maintenance and/or accumulation of organic matter

Organic matter is the basis for soil fertility in agricultural systems because it performs multiple physical, chemical, and biological functions, and is the principal source of nutrients for plants through recycling. For these reasons, and because chemical fertilizers are not used on small farms, soil organic matter takes on a special importance.

Crop residues in conventional systems are not enough to compensate for the loss of organic matter, due to the high rate of mineralization that occurs in tropical and subtropical climates. On small farms, the only practical and economic means to maintain and/or increase soil organic matter levels is to utilize a green manure/cover crop, or a combination of GMCCs, that has a high potential for biomass production in order to complement the contribution of crop residues. The different species contribute different quantities of organic carbon (Table 4).

TABLE 4
Production of biomass (dry matter) and quantity of organic carbon accumulated annually in the above-ground parts of some species of summer and winter green manure/cover crops (average of 3 agricultural periods). Choré Experimental Station.

Green manure/cover crops	Dry Matter (kg/ha)	Organic Carbon	
		% of D.M.[1]	kg/ha
Summer Species[2]			
Pigeon pea	9,153	56.30	5,153
Black-seeded mucuna	7,500	52.15	3,911
Jack bean	7,703	50.15	3,863
Winter Species[3]			
Oilseed radish	4,771	46.52	2,219
White lupine	4,012	47.97	1,925
Black oats	3,680	49.40	1,818
Hairy vetch	2,942	47.43	1,395
Black oats + Hairy vetch	5,440	48.55	2,641
Black oats + White lupine	4,259	48.57	2,069

[1] D.M. = Dry Matter. For summer species, data was taken from tissue analysis from Brazil (Calegari et al., 1991).
[2] Sown in association with corn, and dry matter determined at approximately 7 months.
[3] Sown after cotton, and dry matter determined at approximately 4 months.
Source: Adapted from Derpsch & Florentín, 1992 and Florentín, 2000.

> Green manure/cover crops that have a high potential for biomass production allow soil organic matter levels to be maintained and/or increased in a way that is both practical and economical.

FIGURE 13
The abundance of nodules that form on the roots of legumes is an indicator that nitrogen is being added to the soil.

3.2.3 Addition of nitrogen (biological fixation) and nutrient recycling

In production systems on small farms, where access to mineral fertilizers is difficult, it is possible to supply the necessary nitrogen through the use of green manures of the legume family. These plants are capable of adding great quantities of nitrogen to the soil through biological fixation in the nodules formed by Rhizobium bacteria on their roots (Figure 13). Subsequent crops may be able to take advantage of this nitrogen.

Another important benefit of green manure/cover crops (leguminous or not), when deep-rooted plants are used, is the potential to recycle nitrogen and other nutrients that had been leached (or washed) to the deeper soil layers and are not available to crops. Also, while GMCCs are growing they act to retain part of the nutrients, and prevent them from being washed away.

The quantity and type of nutrients recycled by green manure/cover crops vary according to species, depending primarily on the total production of dry biomass as well as in the concentration of nutrients in the same. An estimation of the quantity of nutrients supplied by the biomass of some green manure/cover crops, as evaluated at the Choré Experimental Station, is presented in Table 5.

TABLE 5
Estimation of the quantity of macronutrients that accumulate annually in the above-ground parts of some species of green manure/cover crops sown on small farms. Choré Experimental Station.

Green Manure	Dry Matter (kg/ha)	Macronutrients In dry matter[1] (kg/ha)			Ratio C/N[1]
		N	P	K	
Summer Species[2]					
Pigeon pea	9.153	240	13	240	22
Jack bean	7.703	246	12	433	16
Black-seeded mucuna	7.500	192	10	108	21
Winter Species[3]					
Oilseed radish	4.771	86	11	156	19
White lupine	4.012	75	5	55	26
Black oats	3.680	48	7	84	39
Hairy vetch	2.942	81	7	81	17

[1] For summer species, data was taken from tissue analysis from Brazil (Calegari et al., 1991).
[2] Sown in association with corn, and dry matter determined at approximately 7 months.
[3] Sown after cotton, and dry matter determined at approximately 4 months.
Source: Adapted from Derpsch & Florentín, 1992 and Florentín, 2000.

> The addition of nitrogen by legumes, and the recycling of nutrients through green manure biomass and crop residues, is essential for the maintenance and improvement of soil fertility.

One great advantage that the recycling of nutrients by green manure/cover crops offers is that they are released slowly to the soil as the dead residues decompose, allowing the plants that follow to make use of them gradually and more efficiently.

3.2.4 Weed control

One of the aspects that farmers most appreciate about the use of green manure/cover crops (GMCCs) is weed suppression. This allows them to save labor used for hoeing and to reduce the use of herbicides, lowering production costs (Figure 14). On small farms, where the possibility of using herbicides is limited due to lack of knowledge about their application and lack of access, GMCCs are essential for the success of Conservation Agriculture.

FIGURE 14
Cotton crop over mucuna cover with very few weeds. The use of green manure/cover crops reduces the cost of, and labor for, weed control.

Work done at the Choré Experimental Station showed that weed infestation was markedly reduced in plots of cotton sown after summer green manure/cover crops associated with corn, in comparison with the plots where GMCCs were not used (Table 6). According to these results, the GMCCs that displayed the best weed control were grey-seeded mucuna and sunnhemp, with reductions in infestation of 95 and 93%, respectively, in relation to the check plot without GMCC that had 2,676 weed plants/m².

The lower population of weeds in the cotton field due to the use of green manure/cover crops, besides favoring the crop by reducing competition for water, light, and nutrients, may represent an additional economic advantage for the farmer by saving the labor (hoeing) for weed control, which often leads to abandonment of the crop (Figure 15).

Winter green manure/cover crops also reduced the incidence of weeds in the following corn crop (Table 7).

FIGURE 15
The overgrowth of weeds in cotton in conventional cultivation systems, especially of sandbur, often leads to abandonment of the harvest.

TABLE 6
Weed infestation in plots of cotton sown after different species of summer green manure/cover crops, and in the check plot without a GMCC (60 days after sowing). Choré Experimental Station.

Treatments	Weeds		Species (%)	
	plants/m²	relative %	Narrow-leaved[1]	Broad-leaved[2]
Without GMCC	2.676	100	19	81
Cowpea var. Tupí	782	29	1	99
Lab-lab[3]	675	25	20	80
Pigeon pea	518	19	15	85
Calopo	425	16	4	96
Jack bean	382	14	3	97
Black-seeded mucuna	300	11	7	93
Sunnhemp	191	7	9	91
Grey-seeded mucuna	130	5	9	91

[1] Predominant weeds: sandbur (Cenchrus echinatus), sourgrass (Digitaria insularis).
[2] Predominant weeds: dayflower (Commelina sp.), painted spurge (Euphorbia heterophylla), hispid starburr (Acanthospermum hispidum).
[3] White seed.
Source: Adapted from Florentín, 1997.

> On small farms, where the possibility of using herbicides is limited due to lack of knowledge of their application and lack of access, green manure/cover crops are essential for the success of Conservation Agriculture.

TABLE 7
Weed infestation after flattening winter green manure/cover crops with a knife-roller and before the sowing of corn. Choré Experimental Station.

Green manure/cover crop	N° of weeds/m²				Dry matter of weeds	
	Narrow-leaved[1]	Broad-leaved[2]	Total	Relative %	g/m²	Relative %
Winter fallow	15	85	296	100	144	100
Sweet white lupine	15	85	264	89	119	83
Peas var. Arvejón	21	79	196	66	105	73
Triticale[4]	18	82	159	54	58	40
Bitter white lupine	23	77	146	49	57	40
Sunflower	6	94	142	48	34	24
Black oats	29	71	99	33	8	6
Oilseed radish	9	91	46	16	8	6
Black oats + Hairy vetch	0	100	30	10	4	3
Hairy vetch	0	0	0	0	0	0

[1] The predominant wide-leaved weed was Brazilian pusley (Richardia brasiliensis).
[2] The predominant narrow-leaved species was Jamaican crabgrass (Digitaria horizontalis).
Source: Adapted from Florentín, 1997.

In the Department of Paraguarí, in the cultivation of No-Till sugar cane, the inclusion of grey-seeded mucuna as cover for the first year, and of dwarf mucuna the second year, reduced the number of hoeings per year to one; the conventional system had to be hoed three times per year. (Néstor Paniagua, personal communication, 2000). Similar results were obtained in the Department of Caaguazú, with the use of grey-seeded mucuna as cover for the cultivation of watermelon (Celestino Portillo, personal communication, 2000).

> Green manure/cover crops, because of their ability to suppress weeds, allow the adoption of the Conservation Agriculture on small farms without the use of herbicides.

Other experiences in the Department of Paraguarí showed that pigeon pea sown at high densities reduced the incidence of difficult-to-control weeds such as nutgrass (*Cyperus rotundus*), and succeeded in eliminating grasses such as Jamaican crabgrass (*Digitaria horizontalis*). Also, grey-seeded mucuna reduced the infestation of sandbur (*Cenchrus equinatus*) on several farms in the Department of San Pedro.

3.2.5 Increase and maintenance of crop productivity and decrease in production costs

When adequately chosen, green manure/cover crops (GMCCs) result in significant increases in the production of the crops that follow. However, in the long term, of interest is not only the increase but also the maintenance of high production. The use of GMCCs is indispensable for the achievement of this objective (see item 3.8).

> The correct selection of green manure/cover crop species to precede a certain crop makes it possible to achieve a higher yield and greater economic return.

Another important aspect of green manure/cover crops is that they make No-Tillage viable through biological tillage and the reduction of weed incidence. And so, with this system, the work of soil preparation (plowing, etc.) is avoided, and crop management practices (weed control, hilling-up, etc.) are reduced. This, added to the possibility of utilizing less fertilizers and other inputs (eventually herbicides and other pesticides), permits a reduction in production costs for those crops that follow GMCCs.

One more advantage of green manure/cover crops is that they require little investment of capital in that, normally, the farmer can produce and sow the seed himself.

3.3 GENERAL REQUIREMENTS THAT SHOULD BE MET BY GREEN MANURE/COVER CROPS

In order to achieve the greatest benefits from green manure/cover crops (GMCCs), it's necessary to know all aspects related to them (family to which belong, growth habit, cycle, rusticity, weed competition, effects on soil, nutrient recycling, nitrogen-fixing or not, seed production, performance when faced by pests and diseases, ways to manage, etc.). Also, it's necessary to know the objectives being sought by including them, as well as aspects concerning the production systems in which they will be included (climate, soil type, conditions of soil fertility, crops with which they will be integrated in the system, available machinery and/or implements, etc.).

As a general rule, the species of green manure/cover crop (GMCC) selected should be rustic and require few crop management practices. Normally, when soils are apt for traditional agricultural crops (cotton or corn) it should not be necessary to fertilize GMCCs, nor to apply lime. An exception would be in those cases where a system fertilization strategy is used, placing fertilizer with the GMCCs and not with the commercial crop. However, when soils are very degraded, the utilization of fertilizers and lime should be considered in order to achieve good growth of the green manure/cover crops and so produce enough biomass to initiate recuperation of those soils.

There exist some **requirements** that green manure/cover crops should meet, that would make them more favorable for incorporation into an agricultural production system, the most important being:

- Have low establishment and management costs.
- Be easy to sow and manage.
- Achieve good shading and weed suppression.
- Produce a favorable residual effect on cash and subsistence crops.
- Be rustic and require few crop management practices.
- Present good conservation characteristics.
- Avoid proliferation of pests and diseases.
- Avoid competition for land, labor, time, and space with cash or subsistence crops.

The principal **conservation characteristics** that green manure/cover crops should have and that should be considered in their selection are the following (Monegat, 1991):

- Rapid growth and good soil cover under prevailing soil and climatic conditions.
- Production of a great quantity of green and dry mass, of the above-arts and roots.
- Slow decomposition of dry matter produced.

3.4 SPECIES OF GREEN MANURE/COVER CROPS MOST ADAPTED TO SMALL FARMING SYSTEMS

Green manure/cover crops (GMCCs) are differentiated by their period of growth: summer GMCCs grow during the spring and summer seasons, and winter GMCCs develop during the fall and winter seasons.

3.4.1 Spring-Summer Species

- **Grey-seeded mucuna** *(Mucuna pruriens = Stizolobium cinereum)*

Characteristics: It is an annual, herbaceous, creeping, climbing legume. It is of medium size, has rapid initial growth, is rustic, tolerant of pests and diseases, and also controls nematodes. It produces a good quantity of dry matter under conditions of medium and high fertility (8 to 10 t/ha); however it does not develop well on extremely degraded soils (2 to 4 t/ha). It possesses a vigorous root system capable of biological fixation of atmospheric nitrogen. It is an excellent green manure/cover crop for the majority of crops that follow it in rotation, and the effect it has on corn, cotton, cassava, tobacco, and vegetables is outstanding. It also has a very marked effect on weed suppression, both during growth by smothering them and as dead cover through shading and allelopathy[1] (Figure 16).

FIGURE 16
Grey-seeded mucuna is a species that produces a great quantity of green and dry matter, protects the soil, and smothers weeds efficiently.

Ways to use: One of the alternatives most practiced is the sowing of mucuna in association with corn (Figure 17).

It is recommended that corn be sown early (August-September), in order to install the mucuna as soon as possible (November-December). The earlier it is sown during this period, the greater the production of biomass. For this system, it is recommended that 2 rows of mucuna be sown between every two rows of corn, 90 to 110 days after the corn was sown, and not before, to prevent the mucuna from smothering it.

FIGURE 17
Grey-seeded mucuna associated with corn is the best way to use this green manure/cover crop on small farms.

---

[1] **Allelopathy:** effect of chemical substances released during the growth or breakdown of certain plants, which impair or impede the germination or development of others.

Recommended spacing is 50 cm between rows and 40 cm between holes, with 1 to 3 seeds per hole. Seed use is approximately 100 to 120 kg/ha. The weight of 1,000 seeds varies from 1,000 to 1,300 gm, depending on the year.

Approximately 40 days after sowing the mucuna (depending on the corn variety), the corn is harvested manually or left among the mucuna until winter. Grain conservation is good in the latter case, with less attack by insects, especially weevils. After the corn is harvested, the mucuna is left to grow until it is time to flatten it. It is recommended that mucuna be flattened with a knife-roller 15 days before sowing the following crop, in order to crush the cover but without giving time for new weeds to grow or for the biomass to degrade. When there are frosts, or the mucuna dies because its cycle has ended, it's generally not necessary to flatten the mucuna. Otherwise, it may be flattened in July-August with a knife-roller, machete, or herbicides. If there are viable seeds, it will be necessary to harvest them in order to prevent them from resowing naturally and becoming weeds in the following crop.

Another option to use grey-seeded mucuna is to sow it following summer crops that are harvested in January-February (corn, peanuts, tobacco). The field often have few weeds and so there is no need for weed control prior to sowing the mucuna. If necessary, weeds can be controlled with a knife-roller, by hoeing or cutting, or with herbicides. The sowing method and density are the same as recommended for the association of mucuna with corn.

> Grey-seeded mucuna is one of the species most utilized in production systems on small farms with soils of medium to high fertility, principally in association with corn.

Between the end of mucuna's cycle and the sowing of crops such as cotton and sesame (October-November), there is a period of about 2 months with no crop. In order to avoid the proliferation of weeds during this period and to obtain a greater production of biomass and cover, it's possible to advance the flattening of mucuna to June and sow a winter green manure/cover crop, preferably a mixture of black oats, white lupine, and oilseed radish.

> The best time to sow grey-seeded mucuna is in November. Sowings of January and later generally produce little biomass.

Seed production: Normally, seed can be harvested from mucuna used as a green manure/cover crop if it was sown early. However, for greater seed production it is recommended that mucuna be sown at a lower density (12 to

15 kg/ha), with support such as bamboo tripods or tree trunks and branches, or in association with agricultural crops such as corn, old cassava, castor, etc. In this way the rot of flowers and seedpods from contact with the soil is avoided (Figure 18). Fences may also be used as support.

One practical system is to use corn as a support, sowing the mucuna around 30 days after planting the corn. In order to avoid possible losses due to frost, mucuna should be sown in October. In this system it is recommended that 1 row of mucuna be sown between every other row of corn, leaving 2m between rows and 90 cm between holes, and depositing 2 seeds per hole (12 to 15 kg/ha of seed). Mucuna's cycle ends in July-August at 240 to 270 days after sowing. Expected yield of seed is from 1,000 to 1,500 kg/ha. It is important to classify seeds, discarding those that are rotten or with black coloration (they won't germinate). A seed-producing area of 1,000 m^2 is needed to produce enough seed to sow 1 ha of green manure/cover crop.

FIGURE 18
Seedpods of grey-seeded mucuna grown with support; this considerably improves the yield and quality of seed.

- **Pigeon pea** *(Cajanus cajan L. Millsp.)*

Characteristics: It is a semi-perennial (2 to 4 years), shrubby, tall legume. It is characterized as being very rustic in terms of fertility and soil type. It tolerates both drought and cold. When it freezes, the leaves dry out but it resprouts immediately. In the case of a hard frost, part or all of the plant may die. It produces great volumes of biomass annually, even on extremely degraded soils (7 to 14 t/ha); for this reason, it is an excellent option to initiate soil recuperation under these conditions (Figure 19).

FIGURE 19
Exuberant growth of pigeon pea on extremely degraded soil in which cassava no longer produces (foreground).

Initial growth is slow, therefore weed infestation may occur. In this case, it is necessary to hoe in order to achieve good initial growth. Later, it covers the soil well and controls weeds with its shade. Besides the species mentioned there exists a dwarf pigeon pea, which is quite a bit shorter and has smaller pods and seeds.

Ways to use: It is recommended principally for the recuperation of degraded soils, grown in association with corn: a) as an alternative to traditional fallow, leaving the plants to grow for 2 to 4 years; or b) as the first GMCC in an annual crop rotation system. These alternatives may also be practiced on more fertile soils. The most appropriate moment to sow pigeon pea is 60 to 70 days after sowing corn, normally after the second hoeing, but it may be moved forward to 30 days to coincide with the first hoeing. It is recommended that 2 rows of pigeon pea be sown between each two rows of corn, placing 4 to 6 seeds per hole at a distance of 30 cm between holes and using 25 to 35 kg/ha of seed. The weight of 1,000 seeds is approximately 140 to 170 g, depending on the variety.

When used in rotation with annual crops, it is recommended that pigeon pea be flattened 2 to 3 weeks before the following crop is sown. This operation may be done with a knife-roller, later cutting with a machete any stems that rise above the soil level so that they don't resprout. This method reduces by half the time used when flattened with only a machete.

> Pigeon pea, because of its rusticity and high production of biomass, is an ideal plant to recuperate extremely degraded soils.

FIGURE 20
Strip of pigeon pea left standing to produce seed (right), a practice recommended so that a farmer may obtain his/her own seed.

Seed production: It is possible to take advantage of the same sowings intended for green manure/cover crop, leaving a strip not flattened (Figure 20). Better seed production is achieved in crops that are less dense; to that effect, it is possible to sow 1 or 2 rows of plants in the form of live fences. The seedpods of the majority of tall varieties mature in July-August, approximately 210 to 270 days after sowing, while those of the dwarf varieties mature in May-June (180 to 210 days after sowing). Seeds may be harvested by collecting the dry pods manually, or by cutting the plants; they are later dried and threshed. Seed yields generally vary from 1,000 to 2,000 kg/ha.

The seeds are highly sensitive to weevils, and attack frequently occurs while the pods are still on the plant. It is recommended that seeds be treated with ash, sand, or lime, mixing them in a container. Also, the seeds may be treated with aluminum phosphide (Phostoxin or Gastoxin) immediately after harvest if they are weeviled, and the lot controlled periodically to detect later attack. A seed-producing area of 300 m² is needed to produce enough seed to sow 1 ha of green manure/cover crop.

- **Jack bean** *(Canavalia ensiformis L. DC)*

Characteristics: It is an annual, herbaceous, erect legume that climbs little, is of medium size, and grows vigorously. It is characterized as being less demanding in terms of soil fertility than grey-seeded mucuna. It adapts very well in all of Paraguay's Eastern Region, even withstanding frosts of low intensity. It is also tolerant to drought. It normally has no diseases and, although occasionally attacked by leaf-chewing insects, succeeds in achieving good cover and producing important quantities of biomass in soils of medium and high fertility (approximately 6 to 7 t/ha of dry matter). In extremely degraded soils it produces 2 to 4 t/ha of dry matter.

Jack bean attains rapid initial soil cover due to the great size of its leaves. It competes very well with weeds, an effect that lasts even when cover is reduced by leaf fall. It possesses a vigorous taproot with an outstanding capacity to decompact soil (biological tillage). In spite of early production of the first seedpods, its vegetative development continues for a long period of time (10 to 12 months) during which it again produces seed.

Ways to use: Due to its growth habit of climbing little, it can be cultivated simultaneously with other annual crops (1 row between each two rows of crop), principally with corn (Figure 21) or with cassava when sown beginning in August. This is an advantage because soil cover is present practically all year, with good weed suppression and nitrogen added by biological fixation. However, in dry years simultaneous cropping has caused a reduction in corn yields in Choré, due to competition for water. Another advantage is that, because of its long cycle and cold tolerance, crops such as cotton and sesame may be sown without a break, immediately after passing the knife-roller.

FIGURE 21
Jack bean, due to its non-climbing growth habit, is a species that may be associated early with corn, preferably with the first hoeing or up to 90 days.

Jack bean may also be sown after the crop's first or second hoeing (30 or 60 to 70 days), or later. In perennial plantations (citrus, yerba mate, etc.) it is very efficient as a green cover crop because it doesn't send out climbing branches and it covers the soil for a prolonged time. In sowings that are simultaneous, or at 30 days, it is recommended that one row be sown between every two crop rows, with holes at 30 cm and 1 or 2 seeds per hole. In this case 70 to

80 kg/ha of seed are used. In later sowings (as of 60 days), it is recommended that 2 rows be sown between every two crop rows, which raises seed use to 140-160 kg/ha. The weight of 1,000 seeds ranges from 1,300 to 1,500 g.

It may be flattened with a machete or knife-roller, immediately or 10 to 15 days before the sowing of summer crops (cotton, corn, etc.). Any weed infestation that occurs later should be adequately controlled.

Seed production: Seeds may be produced in the same field used for a green manure/cover crop. They may also be produced in a pure crop, using a spacing of 1 m between rows and 30 cm between holes. The seeds mature erratically, and should be harvested as they ripen. This occurs from approximately 4 months after sowing until around 7 months after (March-July). Later, the seedpods need to be dried in the sun to facilitate threshing. Seed yields range from 1,000 to 1,500 kg/ha. A seed-producing area of 750 m^2 is needed to harvest enough seed to sow 1 ha of green manure/cover crop.

- **Dwarf Mucuna** *(Mucuna pruriens = Stizolobium deeringianum Bort.)*

FIGURE 22
Dwarf mucuna, because of its non-climbing habit, may be associated early with annual crops such as corn.

Characteristics: It is an annual, herbaceous, medium to short legume, with an erect, determinant, and non-climbing growth habit (Figure 22). Production of dry matter is around 4 t/ha in soils of medium and high fertility. It develops little biomass in extremely degraded soils (2 to 3 t/ha of dry matter). It has good capacity for nitrogen fixation and also promotes the control of soil nematodes. In general, there are no pests or diseases that severely affect its development. Dwarf mucuna is affected by frost, which can cause it to die when intense. It achieves good weed suppression in spite of its short cycle and determinant growth. It provides less soil cover during its development than other species of green manure/cover crops.

Ways to use: Dwarf mucuna can be associated with annual summer crops (corn, cassava, etc.) and also with perennial or semi-perennial crops (yerba mate, sugar cane, citrus, pineapple, and others) to precede winter crops (Figure 23). When associated with annual summer crops, it is recommended that 1 row of dwarf mucuna be sown between every two crop rows, with a manual jab planter or sharpened stick. It may be sown simultaneously or up to approximately 60 days after the corn is sown, in holes 40 cm apart and with 2 to 3 seeds per hole. In simultaneous sowings during dry years there may be competition for water. When associated with perennial or semi-perennial crops it may be sown in October-December with a spacing between holes of 50 x 40 cm, with 2 to 3 seeds per hole (70 to 90 kg/ha). The weight of 1,000 seeds varies from 530 to 750 grams. In general, it is not necessary to flatten dwarf mucuna since its cycle ends in April, which allows a winter crop to be sown directly over its residues.

Seed production: It is possible to harvest seed from the green manure/cover crop. Dwarf mucuna's cycle, from sowing to harvest, is 150 to 180 days. It's possible to reach seed yields of 800 to 1,500 kg/ha. Dwarf mucuna's seedpods rot easily when they remain in contact with the soil and there is rain, and so there may be an important loss of seeds if they are not harvested as soon as they ripen (March-May). A seed-producing area of 500 m^2 is needed to harvest enough seed to sow 1 ha of green manure/cover crop.

FIGURE 23
At the end of its cycle in April, dwarf mucuna makes it possible to sow winter crops over a cover that is free of weeds.

- **Sunnhemp** *(Crotalaria juncea L.)*

Characteristics: It is an annual, tall (over 3m) legume, that is characterized as having great biomass production (7 to 8 t/ha of dry matter when sown in January in association with corn). It also grows well on degraded soils and develops much better than mucuna (Figure 24). Its initial growth is rapid and it has an excellent suppressing effect on weeds. It also stands out for its ability to lower nematode populations.

Ways to use: Sunnhemp has an excellent residual effect on the majority of crops, both summer and winter (vegetables and others). It adapts well to late sowings (January-February) after the harvest of corn, peanuts, cowpeas, etc. It is also advisable to associate it with corn (Figure 25) as an alternative to mucuna, sowing it 60 to 70 days after the crop was sown.

FIGURE 24
Sunnhemp, of excellent biomass production and weed suppression on extremely degraded soils, being flattened with a knife-roller.

FIGURE 25
Sunnhemp at 48 days after sowing, associated with corn (recently harvested), with 2 rows between every two rows of crop.

When associated with corn, sunnhemp may be sown broadcast, incorporating seed into the soil with a hoe (when possible, taking advantage of a weeding). Also, it may be sown in furrows (47 to 48 seeds/m) or in holes (30 cm between rows with 14 seeds per hole) with 2 rows between each two rows of crop. If not associated with a crop, sunnhemp may be sown in the same way, using a knife-roller or a disc harrow (regulated to roll over the ground, cutting as little as possible into the soil) to incorporate the seed when sown broadcast. Sunnhemp is also very appropriate for sowing in the season prior to the establishment and renovation of sugar cane (reduces nematode populations and adds nitrogen). For all systems proposed it is recommended that 40 kg/ha of seed be used. The weight of 1,000 seeds is around 40 to 45 g.

It may be flattened with a knife-roller or machete, immediately before the following crop, preferably leaving no more than a week between flattening of the sunnhemp and the sowing of the crop. This management is most suitable for crops to be planted in August-September (corn, tobacco, melon, watermelon, cassava, etc.), because the sunnhemp is found in its optimum moment to be flattened (flowering – seedpods starting to fill). For crops that are sown later (cotton, etc.), it is recommended that flattening be delayed until just before the plants have viable seed. Sunnhemp may also be left until its cycle ends, as long as seedpods are harvested since the seeds produced could become weeds.

Seed production: Fields sown for green manure/cover crops normally produce a good quantity of seed. If present, carpenter and bumble bees (*Xylocopa sp., Bombus sp.*) pollinate the flowers and assure good yields. If the objective is specifically to produce seed, it is advisable to sow long, narrow strips to favor the activity of the bumble bees. There generally occur attacks of the Bella Moth (*Utetheisa ornatrix*) on seedpods, impairing seed formation and ripening. However, it doesn't cause great losses under the conditions of Paraguay's Eastern Region. Seedpods ripen in June-July, around 180 to 240 days after sowing. It may be harvested manually, cutting branches with seedpods, or the pods themselves, and threshing them later with a stick. Yields vary around 600 to 800 kg/ha, and may reach 1,200 kg/ha or more. A seed-producing area of 700 m² is needed to harvest enough seed to sow 1 ha of green manure/cover crop.

- **Pearl Millet** *(Pennisetum americanum L.)*

Characteristics: It is an annual, erect, tall grass, able to reach a height of around 2 to 3 m. In soils with medium and high fertility it has great potential for biomass production (over 10 t/ha of dry matter), which allows it to contribute a great quantity of organic matter (Figure 26) and recycle large quantities of nutrients, principally potassium. In conditions of low fertility its potential for biomass production falls drastically (around 2 to 3 t/ha of dry matter) and in extremely degraded soils it grows very little.

It grows rapidly and competes very well with weeds. In addition, its great quantity of residue, which decomposes slowly after being flattened, leaves an excellent cover for a long time. It has high drought tolerance, developing in regions with precipitations that start at 200 mm. It can withstand soils that are acidic and saline. It has a high potential to tiller and resprout, and serves as forage for animals.

FIGURE 26
Pearl millet (partially flattened) makes it possible to incorporate great quantities of organic matter into the soil in a short period of time.

Ways to use: It is recommended to precede crops such as soy, cotton, cowpeas, etc. as short-period cover, due to its rapid initial growth and short cycle, principally in the warmer regions of Eastern Paraguay (Central and North). In this case it is sown in September-October. In addition, it is suitable for late sowings (January-February), after the harvest of crops such as peanuts, cowpeas, corn, etc. It may be used for forage (grazed or cut and carry), in which case it should be left to resprout in order to produce enough biomass to cover the soil until the following crop is sown. If pearl millet's cycle ends (whether used as forage or not) the seeds may be harvested; otherwise, they will probably be eaten by birds.

Twenty to 25 kg/ha of seed are used in broadcast sowings, and the seed incorporated into the soil surface with a knife-roller, disc harrow (regulated to roll over the ground, cutting as little as possible into the soil), etc. The weight of 1,000 seeds of the common varieties is around 4 to 10 g. It may be flattened with a knife-roller 2 to 3 weeks before the following crop is sown, complemented by herbicides, hoe, etc. in order to control eventual resprouting. It is recommended that pearl millet be associated with a legume such as pigeon pea, sunnhemp, cowpea, or other.

Seed production: Seeds may be harvested from the fields of green manure/cover crop. In order to achieve good yields, it is essential to take note of the risk that seed may be eaten by birds. It is recommended that seeds be harvested as soon as they ripen (May-June) and dried immediately. The harvest may be manual. Heads with dry seeds may be placed in bags (preferably cloth), rubbed by hand to remove the grains, and later winnowed. Pearl millet's complete cycle is from 120 to 150 days and seed yield may reach about 1,000 kg/ha. A seed-producing area of 250 m^2 is needed to harvest enough seed to sow 1 ha of green manure/cover crop.

- **Forage sorghum** *(Sorghum sp.)*

FIGURE 27
Forage sorghum, like pearl millet, is capable of producing a great quantity of biomass in a short period of time.

Characteristics: It is an annual, erect, tall grass. It has rapid initial growth, and covers and protects the soil, controlling weeds well (Figure 27). In addition, it has an allelopathic effect on weeds. It can produce great quantities of vegetative mass in barely two months of growth. Production of dry matter varies from 5 to 10 t/ha depending on variety used, time of sowing, time of development, and soil fertility. Under good conditions it can produce around 20 t/ha of dry matter. It performed better than pearl millet on very degraded soils in the Department of Paraguarí. In addition to forage sorghum, it is possible to utilize grain sorghum; this also develops well, produces great quantities of mass, and makes it possible to achieve a good grain harvest.

Ways to use: It is used principally as a short-period green manure/cover crop in sowings as of August-September, being able to develop for 50 to 70 days until crops such as cotton, sesame and soybeans are sown. It can also be used in late sowings of January-February after the harvest of corn, cowpeas, peanuts, etc., and left until the next summer crop is sown. In this situation, the seed is normally harvested, and the residues then cut flush with the soil in order to take advantage of it as forage and to favor resprouting. With this a good quantity of biomass and cover can be produced. Also, it may be grazed directly by animals and then left to resprout, taking care that it is not grazed before 35 to 45 days after resprouting (see item 3.7.1).

It is recommended that forage sorghum be sown broadcast using 20 to 25 kg/ha of seed, incorporating them superficially into the soil with a knife-roller or disc harrow (regulated to roll over the ground, cutting as little as possible into the soil). The weight of 1,000 seeds of the variety Sudan Grass is around 15 g. It may also be sown in mixture with other species (see item 3.5).

Management to flatten it is done 2 to 3 weeks before the following crop is sown, with a knife-roller; this may be complemented later with an herbicide, hoe, etc. to eliminate sprouts.

Seed production: Seed may be harvested from fields of green manure/cover crops, regardless of the system used. The harvest may be done by cutting the heads with a knife, sickle or pruning shears, and threshing them later. Seeds mature in May-June, at 120-150 days after sowing. Yields generally surpass 1,000 k/ha of seed. A seed-producing area of 250 m^2 is needed to harvest enough seed to sow 1 ha of green manure/cover crop.

- **Lablab** *(Lablab purpureus L., or Dolichos lablab L.)*

Characteristics: It is a biennial, herbaceous legume of medium size, which has a creeping and climbing growth habit similar to that of grey-seeded mucuna, though it is not as aggressive nor does it produce as much biomass (4 to 6 t/ha of dry matter). It is outstanding for its good growth under drought conditions. It is susceptible to nematodes and may increase their populations in the soil. Its long cycle provides cover for an extended time if there are no hard frosts, being able to perform as a biennial plant. Its forage is of better quality than mucuna's and, above all, it is very suitable for making silage in mixture with grasses. On occasion it may suffer attacks of pests such as the beetle of the South American rootworm (*Diabrotica speciosa*), bean leaf beetle (*Cerotoma sp.*), and aphids. There are varieties with different seed colors (black, brown, and white) that have different lengths of cycle (short, medium, and long, respectively).

Ways to use: It can be used in the same system recommended for grey-seeded mucuna (associated with corn), principally in places that are dry and of lower fertility, as long as there are no nematode problems. One advantage it presents is that, being less frost sensitive, it is possible to have live cover up until the moment it is flattened in order to sow late crops (cotton, sesame, and others).

It is advisable to sow lablab starting from 60 days after the corn was sown, placing 2 rows of lablab between each two rows of crop (Figure 28), leaving 30 cm between holes and 3 to 4 seeds per hole (about 40 to 50 kg/ha). The approximate weight of 1,000 seeds is 260 g for brown-seeded varieties, 250 g for those with black seeds and 230 g for those with white seeds. Management to flatten it may be done 10 to 15 days before the following crop, with a knife-roller or machete, eliminating new sprouts in the same way or with herbicide.

FIGURE 28
Lablab at 48 days after sowing, associated with corn (recently harvested), and with 2 rows between every two rows of crop.

Seed production: Seed may be harvested from the green manure/cover crop. The seedpods of lablab are ready for harvest around July-August, depending less on the date sown and more on the variety (white-seeded varieties mature about 20 days later than those with brown seeds and these, in turn, approximately 20 days later than those with black seeds). This means greater difficulty in producing seed of varieties with lighter seeds in years and places with greater frequency of frost. Lablab has serious storage problems due to weeviling. In addition, it loses its ability to germinate the following year under normal conditions of conservation utilized on small farms, and for that reason seed should be produced each year. Seed yields vary from 500 to 1,000 kg/ha. A seed-producing area of 900 m² is needed to harvest enough seed to sow 1 ha of green manure/cover crop.

- **Forage peanut** *(Arachis pintoi L.)*

Characteristics: It is a perennial, herbaceous, tropical and subtropical legume, creeping but not climbing, and of low growth habit. It displays excellent development on sandy soils, and also on clay soils of medium fertility. Its creeping growth habit is initially slow, but it covers the soil very well once established. It has good biomass production in soils of medium to high fertility (up to about 8 t/ha/year of dry matter) and produces almost no biomass in extremely degraded soils (less than 1 t/ha of dry matter). It generally has no problems with pests or diseases. In addition, it helps to reduce nematodes in the soil. Its sanitary performance is superior to that of grassnut *(Arachis prostrata)*, which may suffer attack by thrips and nematodes.

FIGURE 29
Forage peanut, perennial and with a creeping growth habit, is one of the best green manure/cover crops to use in fields of yerba mate and other perennial species.

Ways to Use: It is normally suggested as a cover crop between rows of perennial crops such as yerba mate (Figure 29), fruit trees, etc.

It is recommended that forage peanut be broadcast sown in October-November, using 8 to 15 kg/ha of seed, and incorporated with an animal drawn disc harrow (regulated to roll over the ground, cutting as little as possible into the soil) or knife-roller. The weight of 1,000 seeds is 161 g. It may also be multiplied by cuttings planted at a distance of 1 to 2 m from each other.

In systems with perennial crops, forage peanut may be left without flattening for several years. In rotation with annual crops it may be flattened with herbicides 10 to 15 days before sowing the following crop, or by opening furrows (with hoe or ripper), or by making narrow paths with a hoe, machete, herbicide, etc. where the crop will be sown. Farmers in Brazil manage peanut by drying it with low doses of herbicide to partially burn the plants, then sow No-Till corn. Afterwards, the peanut resprouts and again covers the soil. It is utilized later as forage or cover.

Seed production: It is possible to obtain seed from fields of green manure/cover crops. It is harvested manually in April-May, completing a cycle of 180 to 240 days. Yield is approximately 500 kg/ha. Harvest is difficult, therefore the price of seed is high. It is recommended that small quantities of seed be purchased and then multiplied in the field. A seed-producing area of 250 m² is needed to harvest enough seed to sow 1 ha of green manure/cover crop.

- **Creeping indigo** *(Indigofera endecaphylla L.)*

Characteristics: It is a perennial legume of indeterminate habit, herbaceous, creeping, and of medium height. It tolerates drought and light frost. When burned by frost, it resprouts when temperatures rise, again covering the soil. It displays good capacity for nodulation, excellent soil cover, and resows naturally. It is outstanding for its ability to compete against weeds, which it smothers. It has no insect or disease problems, however it is recommended that this crop not be sown in areas with nematodes. It grows in clay soils as well as in sandy, acidic soils of low fertility (principally lacking phosphorous). It can produce 5 to 6 t/ha of dry matter.

FIGURE 30
Creeping indigo, a perennial legume, is a good option for green cover in perennial crops.

Ways to use: It adapts well when interplanted with perennial crops such as yerba mate (Figure 30), citrus, grapes, macadamia, mango, etc. In this case, it is recommended that it be sown in the months of October-November. When sown broadcast, 15 to 20 kg/ha of seed are used. The weight of 1,000 seeds is 6 g. Another option is to sow the creeping indigo in rows 50 to 70 cm apart (seeds sown by trickling). In addition, it may be propagated by cuttings. Once established it may be left to grow for several years. Creeping indigo may be managed by drying it with an herbicide, or opening furrows (with a hoe or ripper) through the live cover, or making a narrow path with a hoe, machete, herbicide, etc. to sow a crop.

Seed production: Seeds may be harvested from the fields of green manure/cover crop. They generally reach maturity in September-October, 9 to 12 months after sowing, and may be harvested each year. It is possible to produce 60 to 150 kg/ha or more of seed, which is harvested by hand. A seed-producing area of 3,000 m² is needed to harvest enough seed to sow 1 ha of green manure/cover crop.

- **Tephrosia** *(Tephrosia tunicata L., Tephrosia candida L.)*

Characteristics: It is a perennial, tropical and subtropical legume; it is bushy, has woody stems, is 3 to 4 m tall, and shows slow initial growth. It stands out for its great rusticity, growing well in soils that are clay, acidic sandy, and of low fertility. Biomass production of T*ephrosia candida* varies from 7 to 15 t/ha/year of dry matter. It reduces nematode populations. It has a vigorous taproot, capable of decompacting soil. It may be used to control "vaquitas" (beetles of *Diabrotica spp.*). It has good residual effect, and increases the yields of cash or subsistence crops sown afterwards. It is highly resistant to attack by pests.

FIGURE 31
Tephrosia, a shrub (3 m tall), may be used as an alternative to fallow for the recuperation of degraded soils.

Ways to use: Tephrosia can be utilized in several production systems:

a) For the recuperation of extremely degraded soils, since it can be left for 2 to 5 years, according to the degree of soil degradation (Figure 31). Tephrosia may be sown in rows every 60 cm to 1 m, with 10 to 12 plants per linear meter, or with a manual jab planter with 30 cm between holes and 3-4 seeds per hole. Eight to 15 kg/ha of seed are used. The weight of 1,000 seeds is 15 g for *Tephrosia candida* and 28 g for *T. tunicata*. After the soil has been recuperated, tephrosia can be flattened by cutting the plants flush with the soil and eliminating any sprouts that appear. Because of their woody stems, they are tough to flatten with a machete.

b) As a shrub, it may be used in live fences/vegetative barriers, or in alley cropping. Rows should be separated by 4 to 12 m, principally in areas with steep slopes. Later, it may be cut (pruned) with a machete and the branches and leaves distributed over the soil; afterwards, a crop such as corn, cowpeas, cassava, cotton, peanuts, etc., may be sown. The resprouts are also cut (1 to 2 cuts) and placed between the rows of growing crops.

c) As a cover in association with perennial crops. It may be cut and left between the rows of crops such as yerba mate and others.

Seed production: The cycle is complete in August-September, 240 to 330 days after sowing. The harvest may take place annually during the same period of time. The harvest and threshing are done manually; seed production is from 200 to 400 kg/ha. On average, a seed-producing area of 600 m² (according to minimum yield) is needed to harvest enough seed to sow 1 ha of green manure/cover crop.

- **Leucaena** *(Leucaena leucocephala L. de Wit)*

Characteristics: It is a tree of the legume family, perennial, and of tropical and subtropical climates. When the plants are young (less than 1 year) they cannot withstand frost. Adult plants are affected little by frost; part of the leaves and branches may be burned, but later resprout. It adapts well to a wide range of soils, but doesn't grow well in very acid soil. It is characterized by its great potential for nitrogen fixation and recycling, which may reach over 600 kg/ha/year with 3 to 4 cuts. Furthermore, it has a great capacity to recycle other nutrients, which previously leached (washed), through its deep roots. Like the majority of tree species, its early development is very slow. Its leaves and branches are of high nutritional value for animals, with a high protein content; however, there are limitations for its use as forage (see item 3.7.1).

Ways to use: In agricultural production, leucaena is recommended principally for the recuperation of extremely degraded soils. It is grown in strips of 1 to 2 rows, leaving 6 to 12 m between rows in which annual crops may be grown (Figure 32). With less space between plants (50 cm between holes or trickled into a furrow), a greater quantity of biomass is produced (10 to 12 t/ha of dry matter with rows 6m apart). The weight of 1,000 seeds is around 50 to 60 g. It may be sown in the spring using a manual jab planter (3 to 4 seeds per hole), or seedlings transplanted in autumn (50 cm between holes and 6 m between rows) if a seedbed had been made, or by taking advantage of seedlings from natural regeneration. In order to sow seeds, which are hard when not recently collected, it's necessary to treat them beforehand with hot water to facilitate germination. One option is to pour boiling water over the seeds and stir continually for 2 to 3 minutes. They are then rinsed with cold water to be sown later.

FIGURE 32
Cotton crop in alley-cropping with leucaena (background), showing better development than the cotton without green manure (foreground).

Due to its initial slow growth, for leucaena to establish well it's necessary to control any weeds that emerge after sowing or transplanting. In order to avoid hoeing exclusively for that reason, it is recommended that leucaena be associated with annual crops, principally cassava. In this way, advantage is taken of the crop's weeding while at the same time a suitable microclimate is created for the growth of leucaena. In this system, cassava and leucaena seed are planted simultaneously. In the case of seedlings, it's possible to transplant them later, in the spring.

Leucaena is managed by cutting it periodically (3 to 4 times a year, mainly in spring-summer), at a height of 20 to 50 cm from the soil, before it produces viable seeds that could become weeds. Its fine branches and leaves are placed over the soil surface between rows. As they break down, they are taken advantage of by the crops. The thicker branches may be used as firewood or stakes or, preferably, left along the strips of leucaena where they will decompose. The organic matter that accumulates this way is later redistributed between the rows.

Seed production: Since it is a perennial crop, it is recommended that only 1 or 2 trees be left as a seed bank. The objective would be to replace plants in case of loss, or if there is need to expand the area under the proposed system. These trees should be outside the farmed area in order to avoid weed infestation from natural regeneration. Seed matures in July-August or all year, depending on the variety. It may be harvested annually. It is possible to assure that you will obtain enough seed to sow 1 ha of green manure by leaving 2 or 3 trees for seed production.

3.4.2 Fall-winter species

- **Black Oats** *(Avena strigosa Schreb)*

Characteristics: It is an erect, annual grass of medium height, and very well adapted to the soil and climatic conditions of Paraguay's Eastern Region. It is resistant to attack by rust and aphids, therefore doesn't require any special crop management practices. It produces 4 to 5 t/ha of dry matter in soils of medium and high fertility, but develops little biomass in very degraded soils (1 to 3 t/ha of dry matter). It responds notoriously well to chemical fertilization, principally nitrogenous. In poor soils, even when it produces little biomass, it controls weeds well (allelopathy). Because of its rusticity it surpasses white oats, yellow oats, and rye. There is a variety named IAPAR 61, which has a longer cycle and produces a greater quantity of biomass than the other black oats. It displays a high capacity to tiller, and is able to produce 15 to 35 tillers per plant.

FIGURE 33
Black oats provide a good and persistent soil cover and excellent weed control.

One of the principal benefits of black oats is the excellent soil cover it provides after being flattened, which lasts longer than that of other green manure/cover crops. Shading by the dead cover, added to the very strong allelopathic effect, allows for a high degree of weed suppression; in some situations, it does away with the need for other weed control practices such as hoeing or the use of herbicides (Figure 33).

The use of black oats improves soil health and also promotes increased yields in leguminous crops such as soybeans, cowpeas and Phaseolus beans that follow it in rotation. It is also optimal forage, able to withstand direct grazing in winter and with a good capacity to resprout, which can be taken advantage of as cover.

Ways to use: It is recommended that black oats be sown manually, distributing the seed broadcast with good soil humidity. To favor good germination, it is recommended that an animal-drawn disc harrow (regulated to roll over the ground, cutting as little as possible into the soil), a knife-roller, or a "rake" made of branches be passed after the seed is distributed. The sowing density (broadcast) recommended for pure crops is from 60 to 80 kg/ha of seed for common varieties and 50 kg/ha for the variety IAPAR 61. The weight of 1,000 seeds is from approximately 14 to 20 grams, the variety IAPAR 61 having smaller seed. The recommended sowing time for best plant development is April. However, black oats may be sown from March until the end of winter, which allows it to be incorporated into different production systems. Late sowings run the risk of insufficient rainfall in July and August, and the production of vegetative mass is generally quite reduced.

Black oats may be managed with a knife-roller, flattening the plants 2 to 3 weeks before the following crop. This preferably occurs when the grains are milky, around 120 days for common varieties and 150 to 160 days for the variety IAPAR 61.

Black oats may resprout when flattened early, which would make it necessary to desiccate it with an herbicide, or open furrows (hoe or ripper), or make narrow paths (hoe, machete, herbicide, etc.). On the other hand, when flattened with mature seeds, these may germinate and become weeds in the following crop. It is recommended that black oats be associated with white lupine (see item 3.5) to provide more nitrogen.

> Black oats are outstanding for the excellent soil cover they produce, which lasts longer than that of other green manure/cover crops.

Seed production: It's possible to produce seed in the same field destined for green manure/cover crop. Because of its rusticity, it is exempt from the need for any cultural practices or phytosanitary treatment. On poor soils, black oats need to be fertilized. Seeds mature in August-September, around 140 to 150 days after sowing for common varieties, and 170 to 180 days for the variety IAPAR 61 (September-October). This variety is demanding of cold in the winter in order to achieve good seed production. On small farms the harvest may be manual, either pulling seeds off by hand, or cutting the plant with a machete or sickle to be threshed later with a stick. Approximately 600 to 800 kg of seed may be harvested per hectare. A seed-producing area of 1,200 m² is needed to harvest enough seed to sow 1 ha of green manure/cover crop.

• **White lupine** *(Lupinus albus L.)*

FIGURE 34
White lupine is a legume that is capable of adding approximately 90 kg/ha of nitrogen to the soil, benefiting later crops.

Characteristics: It is an annual, herbaceous, erect legume of medium height. It is an excellent nitrogen fixer by means of bacteria that form nodules on the roots, adding approximately 90 kg/ha of nitrogen (Figure 34). Furthermore, it has a deep tap root system (1 meter or more) that improves the soil's physical condition (through decompaction) and recycles great quantities of nutrients.

White lupine is best adapted to the northern and central regions of Eastern Paraguay, where temperatures are higher and rainfall is less, although it is able to withstand temperatures of -3°C to -4°C. During initial growth it is sensitive to drought, but once its root system is developed it grows well with temporary water deficits. In more humid regions, such as in the Departments of Itapúa and Alto Paraná, or when conditions of high precipitation occur after sowing in Eastern Paraguay's central region (Departments of Paraguarí and Cordillera), it frequently suffers attack by anthracnose, a disease that can kill the plant and limit its cultivation. This problem decreases with crop rotation and, above all, by sowing white lupine in mixture with gramineous species, principally black oats. In this way, it's also possible to compensate for loss of cover caused by the disease. Never harvest plants infected with anthracnose when the grain will be used for seed.

Production of white lupine biomass varies little between soils of different fertility (around 4 t/ha of dry matter) when sown densely and with at least 80 kg/ha of seed.

Ways to use: It is appropriate to precede crops that are demanding of nitrogen, such as corn and cotton. It is recommended that white lupine be sown in April, with a manual jab planter and at a spacing of 50 to 70 cm between rows, 30 to 40 cm between holes, and 3 to 4 seeds per hole (60 to 80 kg/ha of seed). The weight of 1,000 seeds is from 350 to 400 g. A delay in sowing almost always results in an appreciable decrease in growth and biomass

> White lupine, like other leguminous green manures, adds large quantities of nitrogen that improve the yields of later crops such as corn and cotton.

production. When sowing white lupine, it is important to pay attention to seed depth; if seeds are deeper than 4 cm, emergence is difficult and they may not even germinate.

The scarce cover of white lupine during its initial phase of growth allows weeds to proliferate, especially when the distance between rows is wide. To counteract this problem, it is recommended that it be associated with other species such as black oats (see item 3.5), or that sowing density be increased by decreasing the distances between rows and between holes. In some zones, it may be necessary to inoculate seeds with the specific bacteria (*Rhizobium lupini*) to assure or improve nitrogen fixation.

It may be flattened with a knife-roller and/or machete immediately before the following crop.

Seed production: Part of the green manure/cover crop can be left to mature for seed production, in which case it may be necessary to control weeds. It completes its cycle in October-November, approximately 180 days after sowing. It can be harvested by cutting the plant with a machete, and later threshed with static threshers, by beating the dry seedpods with a stick, or simply shelled by hand. Seed production varies from 1,300 to 2,200 kg/ha; one harvest was reported of 3,000 kg/ha on a mechanized farm in the Río Verde Colony (Department of San Pedro). Seed storage under natural, cool conditions (18 to 25 °C) for a year, or in a cold storage room, is the best way to reduce to almost zero the level of anthracnose inoculum (Cardoso & Telardi, 1990). An alternative for small farms could be to dig a hole and bury the seed, properly protected with plastic "canvas". A seed-producing area of 500 m^2 is needed to harvest enough seed to sow 1 ha of green manure/cover crop.

- **Oilseed radish** *(Raphanus sativus L. var. oleiferus Metzg)*

Characteristics: It is an annual plant that belongs to the cruciferous family, is herbaceous, erect, and of medium height. Initial growth is very fast. It's not very rustic, but produces a great quantity of biomass (Figure 35) in conditions of medium and high fertility (up to 5 t/ha of dry matter). Its use is not recommended on extremely degraded soils, since its development is very poor (less than 2 t/ha of dry matter).

Some varieties stand out as having a deep tap root, capable of breaking compacted layers of soil (Figures 36 and 37). It also has the capacity to recycle

FIGURE 35
Oilseed radish produces a great quantity of biomass, controls weeds and has a positive residual effect on subsequent crops.

FIGURE 36
Oilseed radish's taproot is capable of breaking compacted layers of soil, leaving channels after it decomposes.

FIGURE 37
Channels left by decomposing roots of oilseed radish favor water infiltration and penetration by the roots of subsequent crops.

great quantities of nutrients (principally nitrogen) and to make soluble some elements (phosphorous) that cannot be used by plants, making them available for the following crops.

Oilseed radish suppresses weeds very efficiently because of its rapid shading of the soil and its allelopathic effect. The rapid degradation of its biomass right after it is flattened or the end of the plant's cycle, allows rapid weed growth and the occurrence of problems related to low coverage. This situation can be improved by associating oilseed radish with other species of which the biomass decomposes more slowly, for example black oats (see item 3.5).

> The vigorous roots of oilseed radish are capable of loosening the soil and of breaking up compacted layers (biological tillage), favoring development of the crops that follow.

Ways to Use: Oilseed radish shows good residual effect on corn and other crops such as tomato, *Phaseolus beans*, peppers, etc. However, it is normal to observe a reduction in the early growth of corn sown after oilseed radish (from allelopathy). This effect disappears later and has little influence over yield. It is recommended that it be sown in April to May. Late sowings cut short the vegetative cycle (photoperiod) and therefore produce less biomass.

Oilseed radish may be sown manually by broadcasting the seed and incorporating it superficially with an animal drawn disc harrow (regulated to roll over the ground, cutting as little as possible into the soil), a knife-roller, a rake of branches, etc. It germinates well when the seeds are not covered, as long as soil humidity is maintained by drizzle and/or cloudy days, and also when the soil has sufficient dead cover. In pure sowings it is recommended that 20 kg/ha of seed be used, reducing this quantity when sown in association with other green manure/cover crops (see item 3.5). The weight of 1,000 seeds is from 6 to 14 g.

Management to flatten it may be done with a knife-roller, log, machete, etc., 10 to 15 days before the next crop. The optimal moment is when the plants are flowering and the siliques (fruits) begin to fill, about 120 days after sowing. If flattening is delayed and viable seeds remain, these will germinate and may become weeds in the following crop, although they won't develop successfully in hot weather.

Seed production: On small farms it is possible to leave part of the green manure/cover crop for seed production. However, information exists that shows greater yields when lower sowing densities (10 kg/ha) are utilized. Oilseed radish crosses easily with wild radish (*Raphanus raphanistrum*), therefore fields for seed multiplication and neighboring areas should be free of this weed. The seeds mature gradually between 150 and 180 days after sowing. It is recommended that harvest take place in October when the seeds are totally mature (very dry), cutting the plant with a machete in order to thresh the siliques. Yields range from 300 to 500 kg/ha. A seed-producing area of 700 m² is needed to harvest enough seed to sow 1 ha of green manure/cover crop.

- **Hairy vetch** *(Vicia villosa Roth)*

Characteristics: It is an annual, herbaceous, creeping, climbing legume of temperate and subtropical climates. It has the capacity to develop in acid soils with aluminum present. It produces about 3 t/ha of dry matter on soils of medium fertility. It doesn't develop well on extremely degraded soils, producing in this case less than 2 t/ha of dry matter. It displays slow initial development, but because of its growing habit (climbing stems and creeping growth) achieves good soil cover and smothers weeds.

Ways to use: Hairy vetch is preferably sown in April, following summer crops (cotton, corn, sesame). Because of its long cycle (7 to 8 months), it is an interesting alternative to reach October-November with good soil cover (Figure 38) and few weeds. This facilitates No-Till sowing of cotton, sesame, corn, or other late-sown crops. Another alternative is to associate hairy vetch with out-of-season corn or cassava, since it may be sown from April until July. It is preferable that hairy vetch be mixed with black oats or rye (see

FIGURE 38
Hairy vetch, with creeping growth habit and a long cycle, achieves total soil cover for a prolonged period of time (until October).

item 3.5). On small farms, it is recommended that it be sown broadcast, using 50 to 60 kg/ha of seed for pure sowings. The weight of 1,000 seeds is about 25 to 38 g. For good germination, it is necessary that seeds be in contact with the soil; this is achieved by passing a knife-roller, disc harrow (regulated to roll over the ground, cutting as little as possible into the soil), etc.

When one wishes to sow crops over hairy vetch cover, it should be flattened 10 to 15 days before sowing. Management to flatten hairy vetch may be carried out until October, desiccating it with an herbicide, or opening furrows (hoe or ripper), or opening narrow paths (hoe, machete, herbicide, etc.). In these two last cases, the hairy vetch will continue to develop between rows. After October, its cycle will end or it will die from the heat.

Seed production: It is necessary to dedicate a plot exclusively for seed production, because only a small quantity is produced at the sowing densities used for a green manure/cover crop. Furthermore, it has a very long cycle that could delay the sowing of the following crop. In order to produce seed, it is recommended that 25 to 30 kg/ha of seed be used. It is preferable that hairy vetch have a support, which could be a grass such as oats, sorghum, or corn, or an erect legume such as white lupine or pigeon pea. This way, hairy vetch will climb and have more light and air circulation, increasing flowering and yields. Seeds mature in November-December; its cycle is completed at 200 to 240 days. Hairy vetch seedpods have a high incidence of dehiscence (they split open), therefore should be harvested when 60 to 70% of seedpods are dry. A delay in harvest could imply an important loss of seed. Yields may range from 300 to 600 kg/ha in a pure crop, up to 700 to 1,000 kg/ha in a crop with support. A seed-producing area of 2,000 m² is needed to harvest enough seed to sow 1 ha of green manure/cover crop. Seeds should be stored under cool conditions.

- **Rye** *(Secale cereale L.)*

FIGURE 39
Rye has a development similar to that of black oats, and may be an alternate for it in production systems.

Characteristics: It is an annual, erect grass of medium height. Among grains, it is considered to be the species most resistant to cold and frost. It is drought tolerant. It does not withstand waterlogged soil, but develops well in humid soils with good drainage. It's capable of growing on acid soils, either sandy or clay. In soils of medium and high fertility it displays good growth (around 3.5 to 4 t/ha dry matter). It does not develop well on extremely degraded soils where it produces less than 2 t/ha of dry matter. It responds well to chemical and organic fertilization, and competes well with weeds.

Ways to use: In crop rotations it may be sown in April, preferably to precede a leguminous crop, being an alternative to the use of black oat (Figure 39). It adapts well to broadcast sowing; 80 kg/ha of seed are utilized. The weight of 1,000 seeds is from 15 to 18 g. It may be associated with grasses (oats, ryegrass), legumes (vetch, white lupine, etc.), and crucifers (oilseed radish).

It may be flattened to form cover with a knife-roller 2 to 3 weeks before the following crop, preferably while in the milky grain stage, at which time there is a lower incidence of resprouting (around 130 days after sowing). If sprouting should occur, the recommendation is to desiccate it with an herbicide, or open furrows (hoe or ripper), or make narrow paths (hoe, machete, herbicide, etc.).

Seed production: Seeds may be harvested from the same field sown for a green manure/cover crop. It is important to pay attention to the time of harvest (September-October); rye sheds its grain easily, therefore a delay could cause great losses. Productivity may vary from 800 to 1,500 kg/ha. The crop's cycle is complete in about 140 to 160 days. The seed can be harvested and threshed manually. A seed-producing area of 1,000 m^2 is needed to harvest enough seed to sow 1 ha of green manure/cover crop.

- **Ryegrass** *(Lollium multiflorum Lam)*

Characteristics: It is an annual, erect grass of medium height. Although it develops in subtropical climates, it is demanding of cold, therefore its greatest potential for utilization is in the colder regions (Departments of Itapúa and Alto Paraná). It grows well on soils of moderate to high fertility, producing 3.5 to 5 t/ha of dry matter. It doesn't grow well on very degraded soils, where it produces less than 3 t/ha of dry matter. It responds well to the addition of nitrogen and phosphate mineral fertilizers, and also to organic fertilizers. It is demanding in terms of humidity, but does not tolerate being waterlogged. Association with other species, such as black oats, can favor its establishment by creating a favorable microclimate, principally in warmer regions. It demonstrates a high capacity for natural resowing in cold regions, which is an advantage in systems where there are no winter crops, or in soils with steep slope that are left fallow (Figure 40). Seeds that fall in the spring germinate the following autumn. In this case, ryegrass can become a weed in winter crops. The harvest and threshing of ryegrass are done manually.

FIGURE 40
Ryegrass displays excellent cover and natural resowing; it protects soil from erosion, especially on slopes.

Ryegrass provides good soil cover, and has a great quantity of roots that contribute to soil aggregation. It reduces the population of nematodes. It is a forage of high quality that can be grazed directly by animals, or in the form of hay. Its initial growth is slower than that of black oats, and it is used as forage in winter and spring. Ryegrass favors weed suppression, reducing the need for control in the following crop (fewer hoeings and/or less use of herbicides).

Ways to use: It has a positive effect when it precedes the cultivation of soybeans and other legumes. It is recommended that it be sown in March to April, the early sowings being suitable for the colder regions. Also, early sowings are more appropriate for the production of biomass for cover or forage, and late sowings for seed production.

Ryegrass may be sown broadcast or in rows, and the seed should remain on the soil's surface. In the case of broadcast sowing, contact between the seed and soil can be favored by passing a log or knife-roller, taking care that the seeds not be buried. When sowing depth exceeds 1 cm, the seed generally does not germinate. In pure sowings a density of 25-30 kg/ha is recommended, whereas in association with other gramineous (oats, rye) and/or leguminous (vetch, etc.) species, it is recommended that approximately 17 kg/ha be utilized. The weight of 1,000 seeds is from 2 to 3 g.

The optimal time to flatten ryegrass for cover is at 130 to 170 days after sowing, when it is in full flower; however, it should be flattened according to the sowing date of the following crop (2 to 3 weeks before sowing). The knife-roller alone is not enough to kill ryegrass, and it is necessary to complement it with an herbicide, or to open furrows (hoe or ripper), or to open narrow paths (hoe, machete, herbicides, etc.).

Seed production: For seed production it is advisable to sow late, in plots especially intended for that purpose, at similar densities as when sown for a green manure/cover crop. Seeds may also be harvested from fields of GMCC. Inclusively, it may be grazed and seed harvested later. Ryegrass is harvested in November, 1 to 2 weeks after the milky grain stage in order to avoid loss from seed fall. Good seed quality may be achieved by drying immediately after harvest, which can be done manually. On small farms, seeds may be dried by spreading them in the sun and stirring periodically. Yield per hectare can vary around 600 to 800 kg/ha. The complete cycle is about 210 days. A seed-producing area of 500 m² is needed to harvest enough seed to sow 1 ha of green manure/cover crop.

- **Sunflower** *(Helianthus annuus L.)*

Characteristics: It belongs to the composite family. It is an annual, herbaceous, erect plant of medium to tall stature (reaches a height of 1 to over 3 m, depending on the variety). It has a deep, abundant root system. It is drought tolerant (can grow with 250 to 400 mm of rain), and has the capacity to grow almost all year in Paraguay's climatic conditions. It stands out for its rapid initial growth and good shading of the soil. This, together with its pronounced allelopathic effect, makes it one of the most efficient plants for weed suppression (when sown at high densities), inclusively in short periods of time, from 50 to 60 days. Its dead cover decomposes rapidly and, therefore, the following crop should be sown immediately after it is flattened to avoid weed problems. It is one of the winter green manure/cover crop species that produces more biomass in soils of medium and high fertility (4.5 to 7 t/ha of dry matter), principally the varieties Guayakán and Peredovic. It does not develop well on extremely degraded soils (less than 2 t/ha). It recycles a great quantity of nutrients through its vegetative mass, but its residual effect on the following crops is not outstanding in relation to other green manure/cover crops.

Due to its susceptibility to several diseases (Alternaria, Rust, Macrophomina, Sclerotinia, and Phomopsis), sunflower should be inserted in an appropriate crop rotation. Furthermore, it doesn't grow well when sown in the same field year after year (self-incompatibility). The recommendation is to wait a minimum of three years before sowing sunflower in the same place.

Ways to use: It may be included between two crops when a short period of time is available (Figure 41); for example, after out-of-season corn that is harvested in July-August and preceding the cotton crop. Hard frost can damage plants when they are at the height of their growing phase.

In addition, sunflower may be utilized in February-April sowings after summer crops such as corn, cotton, peanuts, etc.

It is recommended that varieties such as Guayakán, Peredovik, Estanzuela, etc. be used. Sunflower may be sown with a manual jab planter, pointed stick, etc., at a density of 20 to 25 kg/ha of seed, spaced at 50 cm between rows with 30 cm between holes and 5 to 6 seeds per hole. Sowing depth should be from 3 to 5 cm. The weight of 1,000 seeds of the variety Peredovic varies from 70 to 120 g. It is preferable that sunflower be associated with other GMCCs, such as black oats and hairy vetch, to improve cover; in this situation greater spacing may be used (see item 3.5).

Management to flatten sunflower may be done immediately before the following crop, with a knife-roller or machete, preferably while in full flower (90 to 120 days after sowing). At this time it reaches maximum biomass production. When used as a short-term GMCC, sunflower may be flattened at 50 to 70 days.

FIGURE 41
Sunflower, because of its rapid growth, is ideal as a short-term green manure/cover crop. It grows all year and is drought tolerant.

Seed production: Fields of green manure/cover crop may be utilized for seed production. Ideally, sunflower is harvested when the achenes (grains) are mature and dry. However, due to the high risk of loss to birds, it's possible to harvest ahead of time if the seeds are immediately dried in the sun to avoid problems with fungi and other diseases that can reduce germination and vigor. It may be harvested manually in September-October. The crop's cycle fluctuates around 150 days, depending on the variety and time of sowing. Seed yields may reach approximately 1,000 kg/ha or more. A seed-producing area of 250 m^2 is needed to harvest enough seed to sow 1 ha of green manure/cover crop.

- **Corn spurrey** *(Spergula arvensis L.)*

Characteristics: It belongs to the Caryophyllaceae family. It is an annual, herbaceous plant of temperate climates that adapts well to Eastern Paraguay, especially to the colder regions. In spite of its low height, it achieves good weed suppression and soil cover.

Corn spurrey produces a great quantity of hard seeds. It resows itself naturally, especially in the colder regions, and therefore may become a weed in the case of winter cropping. Corn spurrey's biomass production is around 3 t/ha in soils of medium and high fertility, and less than 2 t/ha on extremely degraded soils.

FIGURE 42
Corn spurrey (in the photo associated with yerba mate), because of its low growth and good cover, is ideal for association with perennial and annual crops.

Ways to use: It is generally utilized following summer crops. Because of its low, non-climbing growth habit it is ideal to associate with annual crops such as out-of-season corn and cassava, and perennial crops such as yerba mate (Figure 42). The recommendation is to sow it broadcast using 20 to 25 kg/ha of seed, and later pass a knife-roller to favor contact with the soil. When sown in association with cassava, seeds may be incorporated with a hoeing, taking care that they remain at a shallow depth (1 to 2 cm) since they are very small. The weight of 1,000 seeds is about 1 to 1.5 g. In general, there is no need to flatten corn spurrey, only wait for its cycle to end. This way it will reseed itself naturally the following autumn. In the event that an early crop is to be sown, corn spurrey may be flattened before the end of its cycle (in the flowering stage, 60 to 90 days after sowing) with a knife-roller, herbicide, etc.

Seed production: Seed may be harvested from the green manure/cover crop. Due to its long flowering period, seeds ripen gradually. Furthermore, mature seeds fall easily from the fruits. Therefore, the harvest must be timed carefully or seeds will remain on the field that could eventually become weeds. Seed is harvested manually in September-October. Seed production may vary between 200 and 500 kg/ha, sometimes more. The complete cycle is approximately 140 days. A seed-producing area of 1,000 m² is needed to harvest enough seed to sow 1 ha of green manure/cover crop.

3.4.3 Summary of characteristics and recommendations for use of green manure/cover crops on small farms

Tables 8 and 9 summarize the characteristics and recommendations for use of summer and winter green manure/cover crops, respectively. For the recommended sowing density it is assumed that seeds have a germination rate of 100%. If the percentage is lower, sowing density should be corrected using the following formula:

$$A = \frac{B \times 100}{C}$$

A: Corrected sowing density
B: Recommended sowing density
C: Germination percentage

TABLE 8
Characteristics of summer green manure/cover crops and recommendations for their use on small farms.

Common and scientific name	Production of biomass (dry matter, t/ha)		Principal uses in agricultural systems					Seed production			Area (m²) to be sown to produce seed for 1 ha of green manure
	Soils of medium and high fertility	Very degraded soils	Ways to include	Timing and method of sowing[1]	Sowing density (kg/ha)[2]	Time to flatten with knife-roller	Weight of 1,000 seeds (g)	Month of harvest and cycle to maturity (days)	Yield (kg/ha)	
Grey-seeded Mucuna (*Mucuna pruriens* = *Stizolobium cinereum*)	8 to 10	2 to 4	Associated with corn 90 to 110 days after it was sown	November-December; in holes (50 x 40 cm, 1 to 3 seeds/hole)	90 to 120	Until July-August: 10 to 15 days before the following crop is sown. It is not necessary later.	1,000 to 1,300	July-August; 240 to 270	1,000 to 1,500	1,000
			Following summer crops that are harvested early	January-February; in holes (50 x 40 cm)						
Pigeon pea (*Cajanus cajan* L. Millsp.)	7 to 14	7 to 14	To recuperate degraded soils, associated with corn, managed annually or left for several years	October-December; in holes (50 x 30 cm, 4 to 6 seeds/hole)	25 to 35	Managed annually: 10 to 15 days before the following crop	140 to 170	July-August; 210 to 270 (tall varieties)	1,000 to 2,000	300
						Managed by leaving for several years: leave to grow freely or cut in the spring - summer		May-June; 180 to 210 (short varieties)		
Jack bean (*Canavalia ensiformis* L. DC)	6 to 7	2 to 4	Associated with corn and cassava, sown simultaneously or up to 60 days after the crop is sown[4]	October-December: in holes (1 row between every 2 crop rows, 30 cm between holes, 1 to 2 seeds/hole)	70 to 80	10 to 15 days before summer crops are sown	1,300 to 1,500	From March until July (ripens gradually); 120 to 210	1,000 to 1,500	750
			Following summer crops that are harvested early	January-February; in holes (50 x 30 cm, 1 to 2 seeds/hole)	140 to 160					

[1] In broadcast sowings it is recommended that seeds be incorporated with a knife-roller or other implement. Sowing in holes is done with a manual jab planter, pointed stick, hoe, etc.
[2] For seeds with a germination rate of 100%.
[3] October-December sowings.
[4] After 60 days it is best to sow 2 rows of jack bean between every two rows of crop (140 to 160 kg/ha).

GREEN MANURE/COVER CROPS (GMCCS)

Table 8 (continued)

Common and scientific name	Production of biomass (dry matter, t/ha)		Principal uses in agricultural systems				Seed production			Area (m²) to be sown to produce seed for 1 ha of green manure
	Soils of medium and high fertility	Very degraded soils	Ways to include	Timing and method of sowing[1]	Sowing density (kg/ha)[2]	Time to flatten with knife-roller	Weight of 1,000 seeds (g)	Month of harvest and cycle to maturity (days)	Yield (kg/ha)	
Dwarf Mucuna (*Mucuna pruriens* = *Stizolobium deeringianum* Bort)	4	2 to 3	Associated with annual summer crops, simultaneously or up to 60 days after the crop is sown	October-December: in holes (1 row between every 2 crop rows, 40 cm between holes, 2 to 3 seeds/hole)	30 to 50	March-May	530 to 750	March-May; 150 to 180	800 to 1,500	500
			Associated with perennial or semi-perennial crops to precede winter crops	October-December: in holes (50 x 40 cm, 2 to 3 seeds/hole)	70 to 90					1,000
Sunnhemp (*Crotalaria juncea* L.)	7 to 8	7 to 8	Associated with corn, sown 60 to 70 days after the crop is sown	October-December; broadcast, or in 2 rows between every 2 crops rows, in furrows (47 to 48 seeds/m) or in holes (30 cm apart, 8 seeds/hole)	40	Immediately before the following crop, preferably before its seeds are viable	40 to 45	June-July; 180 to 240	600 to 1,200	700
			Following summer crops that are harvested early	January-February; in the same way as when associated with corn						

[1] In broadcast sowings it is recommended that seeds be incorporated with a knife-roller or other implement. Sowing in holes is done with a manual jab planter, pointed stick, hoe, etc.
[2] For seeds with a germination rate of 100%.
[3] October-December sowings.

Vol. 12–2010 | 47

Table 8 (Continued)

Common and scientific name	Production of biomass (dry matter, t/ha)		Principal uses in agricultural systems				Seed production			Area (m²) to be sown to produce seed for 1 ha of green manure
	Soils of medium and high fertility	Very degraded soils	Ways to include	Timing and method of sowing[1]	Sowing density (kg/ha)[2]	Time to flatten with knife-roller	Weight of 1,000 seeds (g)	Month of harvest and cycle to maturity (days)	Yield (kg/ha)	
Pearl Millet (*Pennisetum americanum* L.)	10	2 to 3	As short-period cover preceding summer cash crops	September-October; broadcast	20 to 25		4 to 10	May-June[4]; 120 to 150	1,000	250
			Following summer crops that are harvested early	January-February; broadcast		2 to 3 weeks before the following crop is sown				
Forage Sorghum (*Sorghum bicolor* L. Moench)	5 to 10 up to 20	2 to 3	As short-period green manure preceding summer cash crops	August-September; broadcast	20 to 25		15	May-June[4]; 120 to 150	1,000	250
			Following summer crops that are harvested early	January-February; broadcast		2 to 3 weeks before the following crop is sown				
Lablab Lablab purpureum L. Sweet	4 to 6	4 to 6	Associated with corn 60 days after the crop is sown	November-December; in holes (50 x 30 cm, 3 to 4 seeds/hole)	40 to 50	10 to 15 days before the following crop	Brown seed: 260; Black seed: 250; White seed: 230	July-August; 240 to 270	500 to 1,000	900

[1] In broadcast sowings it is recommended that seeds be incorporated with a knife-roller or other implement. Sowing in holes is done with a manual jab planter, pointed stick, hoe, etc.
[2] For seeds with a germination rate of 100%.
[3] October-December sowings.
[4] January-February sowings.

Table 8 (Continued)

Common and scientific name	Production of biomass (dry matter, t/ha)		Principal uses in agricultural systems				Seed production			Area (m²) to be sown to produce seed for 1 ha of green manure
	Soils of medium and high fertility	Very degraded soils	Ways to include	Timing and method of sowing[1]	Sowing density (kg/ha)[2]	Time to flatten with knife-roller	Weight of 1,000 seeds (g)	Month of harvest and cycle to maturity (days)	Yield (kg/ha)	
Forage peanut Arachis pintoi L.	Around 8	Less than 1	Interplanted with perennial crops	October-November; broadcast	8 to 15	Left several years without managing, or can be managed 10 to 15 days before a crop is sown between the rows of permanent crops	161	April-May; annually	500	250
Creeping indigo Indigofera endecaphyla L.	5 to 6	5 to 6	Interplanted with perennial crops	October-November; broadcast or trickled into rows (50 to 70 cm between rows)	15 to 20	Left several years without managing, or can be managed 10 to 15 days before a crop is sown between the rows of permanent crops	6	September-October; annually	60 to 150	3,000
Tephrosia Tephrosia tunicata L.	7 to 15	7 to 15	To recuperate degraded soils, leaving it for 2 to 5 years	In rows 60 cm to 1 m apart, with 10 to 12 plants per linear meter, or with manual jab planter with 30 cm between holes and 3 to 4 seeds/hole	8 to 15		T. tunicata: 28	August-September; annually	200 to 400	600
			In live fences, vegetative barriers 4 to 12 m apart or alley-cropping	In rows	0.7 to 2.3	Left to grow freely or cut in the spring-summer, 2 or 3 times a year (in alley cropping always cut)				
Tephrosia candida L.			Associated with perennial crops	In rows 60 cm to 1 m apart, with 10 to 12 plants per linear meter, or with manual jab planter with 30 cm between holes and 3 to 4 seeds/hole	8 to 15		T. candida: 15			

[1] In broadcast sowings it is recommended that seeds be incorporated with a knife-roller or other implement. Sowing in holes is done with a manual jab planter, pointed stick, hoe, etc.
[2] For seeds with a germination rate of 100%.
[3] October-December sowings.

Table 8 (*Continued*)

Common and scientific name	Production of biomass (dry matter, t/ha)		Principal uses in agricultural systems					Seed production			Area (m²) to be sown to produce seed for 1 ha of green manure
	Soils of medium and high fertility	Very degraded soils	Ways to include	Timing and method of sowing[1]	Sowing density (kg/ha)[2]	Time to flatten with knife-roller		Weight of 1,000 seeds (g)	Month of harvest and cycle to maturity (days)	Yield (kg/ha)	
Leucaena *Leucaena leucocephala* L.	10 to 12	10 to 12	To recuperate degraded soils, leaving it to grow for several years	In strips of 1 or 2 rows (50 cm between holes, 3 to 4 seeds/hole or trickled into a furrow), leaving 6 to 12 m between strips, in which annual crops may be sown	0.3 to 0.7	3 to 4 cuts per year, principally in spring-summer		50 to 60	July-August or all year (according to variety); annually	Around 500	2 to 3 trees

[1] In broadcast sowings it is recommended that seeds be incorporated with a knife-roller or other implement. Sowing in holes is done with a manual jab planter, pointed stick, hoe, etc.

[2] For seeds with a germination rate of 100%.

TABLE 9
Characteristics of winter green manure/cover crops and recommendations for their use on small farms.

Common and scientific name	Production of biomass (dry matter, t/ha)		Principal uses in agricultural systems				Seed production			Area (m²) to be sown to produce seed for 1 ha of green manure
	Soils of medium and high fertility	Very degraded soils[1]	Ways to include	Timing and method of sowing[2]	Sowing density (kg/ha)[3]	Time to flatten with knife-roller	Weight of 1,000 seeds (g)	Month of harvest and cycle to maturity (days)[4]	Yield (kg/ha)	
Black oats (*Avena strigosa* Schreb)	4 to 5	1 to 3	Following in-season summer crops	April; broadcast	60 to 80	2 to 3 weeks before the following crop, preferably when the grains are milky	14 to 20	August-September; 140 to 150 (common varieties) September-October; 170 to 180 (variety IAPAR-61)	600 to 800	1,200
			Associated with or following out-of-season corn or cassava	April-July; broadcast						
White lupine (*Lupinus albus* L.)	Around 4	Around 4	Following in-season summer crops	April; in holes (50 to 70 x 30 cm, 4 seeds/hole)	60 to 80	Immediately before the following crop	350 to 400	October-November; 180	1,300 to 2,200	500
Oilseed radish (*Raphanus sativus* L. var. Oleiferus Metzg)	Up to 5	Less than 2	Following in-season summer crops	April; broadcast	20	10 to 15 days before the following crop, preferably when in flower up to when the fruits (siliques) begin to fill	6 to 14	October; 180	300 to 500	700
			Associated with or following out-of-season corn or cassava	April-July; broadcast						
Hairy vetch (*Vicia villosa* Roth)	Around 3	Less than 2	Following in-season summer crops or associated with out-of-season corn or cassava	April-July; broadcast	60	10 to 15 days before the following crop (up to October)	38 to 40	November-December; 200 to 240	300 to 600, with support: 700 to 1,000	2,000

[1] It is recommended that winter species not be used until after a 2 year period of soil recuperation with summer green manures, with the exception of the white lupine + black oats mixture.
[2] In broadcast sowings it is recommended that seeds be incorporated with a knife-roller or other implement. Sowing in holes is done with a manual jab planter, pointed stick, hoe, etc.
[3] For seeds with a germination rate of 100%.
[4] April-May sowings.

GREEN MANURE/COVER CROPS AND CROP ROTATION IN CONSERVATION AGRICULTURE ON SMALL FARMS

Table 9 (Continued)

Common and scientific name	Production of biomass (dry matter, t/ha)		Principal uses in agricultural systems				Seed production			Area (m²) to be sown to produce seed for 1 ha of green manure
	Soils of medium and high fertility	Very degraded soils[1]	Ways to include	Timing and method of sowing[2]	Sowing density (kg/ha)[3]	Time to flatten with knife-roller	Weight of 1,000 seeds (g)	Month of harvest and cycle to maturity (days)[4]	Yield (kg/ha)	
Rye (Secale cereale L.)	3.5 to 4	Less than 2	Following in-season summer crops	April; broadcast	80	2 to 3 weeks before the following crop, preferably when the grains are milky	15 to 18	September-October; 150 to 160	800 to 1,500	1,000
Rye grass (Lolium multiflorum Lam)	3.5 to 5	Less than 3	Following in-season summer crops in colder zones (Itapúa and Alto Paraná)	March-April; broadcast	25 to 30	2 to 3 weeks before the following crop, preferably when in full flower	2 to 3	November; 210	600 to 800	460
Sunflower (Helianthus annuus L.)	4.5	Less than 2	Following in-season summer crops	February-April; in holes (50 x 30 cm, 5 to 6 seeds/hole)	20 to 25	Immediately before the following crop is sown, preferably in full flower (90 to 120 days)	70 to 120	September-October; 150	1,000	230
			As short period green manure preceding summer cash crops	August-September; in holes (50 x 30 cm, 5 to 6 seeds/hole)		Immediately before the following crop is sown (50 to 70 days)				
Corn spurrey (Espergula arvensis L.)	Around 3	Less than 2	Following in-season summer crops	April; broadcast	20	10 to 15 days before the following crop	1	September-October; 140	200 to 500	1,000
			Associated with crops such as out-of-season corn and cassava.	April-July; broadcast		September-October				

[1] It is recommended that winter species not be used until after a 2 year period of soil recuperation with summer green manures, with the exception of the white lupin + black oats mixture.
[2] In broadcast sowings it is recommended that seeds be incorporated with a knife-roller or other implement. Sowing in holes is done with a manual jab planter, pointed stick, hoe, etc.
[3] For seeds with a germination rate of 100%.
[4] April-May sowings....

Integrated Crop Management

3.5 ASSOCIATION (MIXTURES) OF GREEN MANURE/COVER CROPS

In general terms, mixtures of green manure/cover crops (GMCCs) always present advantages in relation to single or pure crops. The association of green manure/cover crops aims to achieve the following objectives:

- Assure the development of at least one of the species that belong to the mixture, under variable climatic conditions (rainy or dry years, unfavorable temperatures, etc.) that could damage any of the species (growth, diseases, pests).
- Achieve greater production of biomass and/or cover by utilizing species with different growth habits and potential.
- Improve the quality of residues with regard to speed of decomposition and addition of nitrogen and other nutrients, striving for persistent coverage and release of nitrogen.
- Healthier green manure/cover crop. In terms of health, white lupine has benefited greatly when associated with black oats, reducing notably the attack of anthracnose.
- Produce a microclimate that permits the development of species that are difficult to establish in pure sowings (ryegrass in mixture with black oats).
- Improve production and harvest of seed of species with crawling growth habit (hairy vetch associated with grasses, etc.).
- Increase the period of time that soil is covered by including species that have long cycles. For example, hairy vetch, when associated with black oats, keeps growing after the latter has ended its cycle. In this way, good cover may be achieved from April to October.
- Recycle nutrients found at different soil depths, taking advantage of different root systems.
- Improve the physical conditions of compacted soils by combining species with taproots (oilseed radish, legumes) and those with fascicular roots (grasses).
- Weed suppression (greater efficiency through better cover and allelopathy). Examples: White lupine alone is not efficient at weed control. However, in mixture with black oats it achieves excellent control. At the Choré Experimental Station, on a field highly infested with weeds, principally Dayflower (Commelina sp.), a No-Till sowing of black oats was done in April (broadcast, seeds covered with a knife-roller) and white lupine (with manual jab planter), without hoeing nor utilizing herbicides. The green manure/cover crop grew, smothering the weeds, and developed a good quantity of biomass (Figure 43) that left an excellent soil cover for the following No-Till sowing.

One mixture that stands out in conditions of medium to high fertility is the triple association of black oats, white lupine, and oilseed radish (Figure 44). In this case, greater production of biomass and soil cover is established

FIGURE 43
The association of black oats and white lupine provides excellent cover and weed control for the No-Till sowing of different crops..

FIGURE 44
The mixture of black oats, white lupine, and oilseed radish is one of the most recommended, because it guarantees a greater quantity of high quality biomass than from pure crops.

in a cycle of approximately 120 days. In addition to good weed control, the field remains free of weeds for No-Till sowing of subsequent crops, with no need for chemical control. Furthermore, it is a mixture that has good residual effect on crops such as cotton, corn, cassava, sesame, and others; no drawbacks have been observed due to lack of nitrogen (microbial fixation), allelopathy, etc.

For crops that are sown in October-November such as cotton, sesame, corn, etc., an association of GMCCs that includes a long cycle species, such as hairy vetch, is a good alternative to extend the period of soil cover. Mixtures of hairy vetch with black oats, white lupine, rye, etc., produce a stable cover over a longer period of time. The principal mixtures of GMCCs recommended for small farms are presented in Table 10.

The use of mixtures is also recommended in production systems in which the implementation of No-Tillage is being initiated. The reasons are: a) good soil cover is needed, and b) the addition of a large quantity of nitrogen is needed. Moreover, it should be taken into consideration that chemical fertilization is generally not used on small farms. This is particularly important when the following crop is nitrogen demanding (cotton, corn, etc.).

> The use of mixtures of green manure/cover crops from different families is more efficient than sowing the same species alone, because it assures more abundant cover, greater weed suppression, and a balanced addition and release of nutrients, among other things.

3.6. WAYS TO INCLUDE GREEN MANURE/COVER CROPS IN SMALL FARM PRODUCTION SYSTEMS

The key to success in the use of green manure/cover crops is to find the most appropriate ways to include them in production systems on small farms in each of the different regions.

TABLE 10
Principal mixtures of green manure/cover crops recommended for small farms

Species	Density (kg/ha)	Sowing method¹	Distribution (cm)
Black oats + White lupine	50 + 80 to 100	broadcast holes	50 x 30
Black oats + White lupine	50 + 60	broadcast holes	70 x 30
Black oats + Oilseed radish	50 + 5 to 10	broadcast broadcast	
Black oats + Hairy vetch	50 + 40	broadcast broadcast	
Black oats + Ryegrass	40 + 10	broadcast broadcast	
Black oats + White lupine + Oilseed radish	35 + 60 + 5	broadcast holes broadcast	70 x 30
Black oats + Sunflower + Hairy vetch	35 + 20 + 30	broadcast holes broadcast	70 x 30
Pearl millet or forage sorghum + Pigeon pea	10 + 20 to 25	broadcast holes	50 x 30
Pearl millet or forage sorghum + Sunnhemp	6 to 8 + 15	broadcast + rows or broadcast	
Pearl millet + Oilseed radish	10 + 6 to 8	broadcast broadcast	
Pearl millet + Forage sorghum	10 + 6 to 8	broadcast broadcast	

¹ In broadcast sowings it is recommended that the seeds be incorporated with a knife-roller or other implement. Sowing in holes is done with a manual jab planter, pointed stick, etc. When sowing in rows, these may be opened with a ripper, light blade hoe, etc.

3.6.1 Green manure/cover crops (GMCCs) associated with annual crops

In this system, green manure/cover crops (GMCCs) are sown between the rows of annual crops. This presents the advantage that they are installed in a field that is already cultivated, not one prepared exclusively for that purpose. Furthermore, it takes advantage of weeding done for the cash crop to clean the GMCCs. On the other hand, time is gained by covering the soil earlier, reducing the period of time with low soil cover that normally occurs in the early stages of the cover crop's development (30 to 45 days). In that way, weed infestation is avoided.

Corn is one of the crops that offer the greatest potential to disseminate the use of green manure/cover crops in No-Tillage, since this crop is widespread on small farms and performs well with several species of both summer and winter GMCCs.

- **Corn associated with summer green manure/cover crops (GMCCs)**

The sowing of GMCCs in corn may vary according to the species utilized, the objectives sought, and time of sowing.

- Early sowings (simultaneously or after the corn is first hoed, up to 60 days after the corn was sown).

In this system, species that climb little (have limited growth) such as jack bean, dwarf mucuna, etc. may be used, sowing one row of green manurecover crop between every two rows of corn. One of the advantages of this system is that it can be established utilizing a small quantity of seed (1 row between every two crop rows). Furthermore, this facilitates the crop management tasks of sowing and eventual hoeing. On the other hand, the earlier the green manure/cover crops are sown, the greater the soil protection obtained and, probably, two hoeings of corn can be saved. However, experiences at the Choré Experimental Station have shown a reduction of approximately 15% in corn yields from simultaneous sowings with GMCCs, when periods of drought occurred during the corn's vegetative period. This can be avoided by sowing jack bean and dwarf mucuna a few weeks after the corn is sown (with the first hoeing).

Some species of GMCC that climb, such as grey-seeded mucuna and others, may be sown simultaneously with corn as long as the tendrils are managed continuously (approximately once a week). This increases the need for labor in order to carry out this operation; however, it reduces the labor needed for hoeing.

Dwarf mucuna sown in this system finishes its cycle early (April-May), making it possible to sow a winter crop or GMCC, such as peas, vegetables, black oats, white lupine, etc., immediately after it reaches maturity.

- Late sowings (after the second or third hoeing of corn, at approximately 60 to 120 days after the corn was sown).

Species with slow initial growth (pigeon pea), erect habit (sunnhemp, pearl millet), that climb little (jack bean) or are climbers sown on degraded soils (grey-seeded mucuna, lablab), may be sown as of 60 days after the corn was sown. In this case, 2 rows of green manure/cover crop are sown between every two rows of corn in order to guarantee good vegetative cover and biomass.

It is recommended that species of green manure/cover crop that climb (grey-seeded mucuna and others) be sown at the end of corn's cycle (90 to 120 days after sowing) on soils of medium and high fertility. This makes it possible to harvest the corn before it is totally covered by the green manure/cover crop (Figure 45).

FIGURE 45
Grey-seeded mucuna, sown around 3 months after the corn was sown, covers the soil producing a great quantity of biomass and smothering weeds.

- **Corn associated with winter green manure/cover crops (GMCCs)**

In crops of corn that were sown late (out-of-season) and that mature in autumn-winter, it is possible to associate pure or mixed winter GMCCs as of April with the last hoeing. Small-seeded cover crops (black oats, hairy vetch, oilseed radish) may be sown immediately before hoeing in order to incorporate the seeds. Otherwise, this operation should be done with an additional hoeing. If the corn has already been harvested, seed may be incorporated with a knife-roller or animal-drawn disc harrow (regulated to roll over the ground, cutting as little as possible into the soil). In this case, the field is generally infested with weeds therefore it is necessary to control them (with a hoeing, herbicide, etc.) before sowing.

- **Cassava associated with summer and winter green manure/cover crops (GMCCs)**

FIGURE 46
Cassava may be associated with jack bean, which has a non-climbing growth habit, and that achieves good cover and weed control between the rows during the initial stage of the crop's development.

Cassava may be associated with green manure/cover crops (GMCCs) in its initial stage of growth. When planted in July-August, winter GMCCs such as black oats and hairy vetch may be sown between the rows. In plantings from September on, cassava may be associated with summer GMCCs such as jack bean, dwarf mucuna, dwarf and common pigeon pea (with pruning), forage peanut, etc. Another option is to plant the cassava in double rows 80 cm apart, leaving 120 cm between the double rows where 1 or 2 rows of jack bean or dwarf mucuna may be sown.

In cassava crops that were used to produce "seed" (the stems were cut back), and in which the roots will be harvested in the following months, winter GMCCs such as black oats, hairy vetch, white lupine, etc. may be sown between the rows. In this case, the seeds may be incorporated with a hoeing. Later, a summer crop may be sown No-Till over the cover of winter green manure/cover crop.

3.6.2 Green manure/cover crops (GMCCs) in succession with annual crops

On small farms, there generally occur periods of time with no crops during which GMCCs may be included:

1) **January-February:** Some crops such as corn, peanuts, cowpeas, tobacco, etc. are harvested before mid-February, and the majority of farmers don't use these fields until the following spring-summer. In this case, heavy weed

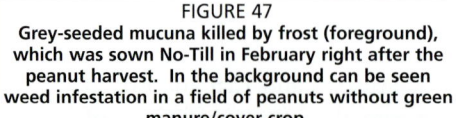

FIGURE 47
Grey-seeded mucuna killed by frost (foreground), which was sown No-Till in February right after the peanut harvest. In the background can be seen weed infestation in a field of peanuts without green manure/cover crop.

infestation occurs. In order to avoid this situation, summer green manure/cover crops that have rapid initial growth may be sown immediately after these crops are harvested. The principal species are: grey-seeded mucuna (Figure 47), pearl millet, lablab, jack bean, sunnhemp, forage sorghum, etc.

After summer crops are harvested, the land often remains relatively free of weeds and there is no need to hoe, or to carry out any additional cleaning, in order to sow the green manure/cover crops. However, when there is heavy weed infestation, or when the sowing of GMCC is delayed, the field should be hoed or herbicides used.

2) **April-May:** In this case winter green manure/cover crops may be sown immediately after the harvest of crops such as cotton, corn, sesame, etc. The most promising species are: black oats, white lupine, hairy vetch, oilseed radish, etc., in pure sowings or, preferably, in mixtures.

FIGURE 48
After cotton residues were destroyed by machete in April, black oats were broadcast sown and are being covered with a knife-roller; white lupine will be sown immediately afterward with a manual jab planter.

Winter green manure/cover crops may be sown after cotton, either before or after the remains of the cotton plants are cut with a machete to destroy them. A knife-roller is later passed to crush the residues and incorporate the GMCC seed (Figure 48). If competitive weeds are present, they may be controlled with an herbicide applied before the cotton is flattened (backpack sprayer) or immediately after passing the knife-roller (backpack, or human- or animal-drawn sprayer). The sowing of green manure/cover crops immediately after the cotton harvest reduces weed infestation, often avoiding the need to control them by hoeing or with herbicides.

When the crop to follow the winter green manure/cover crop will be sown in October or later, it is recommended that long-cycle GMCCs (pure or in mixture), such as black oats IAPAR 61 or hairy vetch, be sown in order to guarantee good soil cover and weed control up to that time.

3) **July-September:** During this time, green manure/cover crops may be sown to precede crops such as cotton, corn, sesame, etc. that are sown in October or later. During this period, fields may remain free of crops and/or

GMCCs in different circumstances: a) after out-of-season corn, which is harvested in June-August (Figure 49); b) when winter GMCCs are sown early (April-May) and unable to cover the time between crops (until the next crop is sown); and c) when grey-seeded mucuna is grown and early frost causes premature death. In all of these cases winter GMCCs (black oats, oilseed radish, sunflower) may be sown in order to obtain good cover and weed control until the summer crop is sown.

FIGURE 49
Broadcast sowing of winter green manure/cover crops in July, among the remains of out-of-season corn, which will be flattened with a knife-roller to favor the seeds' germination

3.6.3 Green manure/cover crops (GMCCs) in association with perennial crops

Summer as well as winter green manure/cover crops (GMCCs) may be sown between the rows of perennial crops such as yerba mate, citrus, pineapple, etc., taking care that the species chosen do not compete with the cultivated plants. Promising summer species recommended for this objective are forage peanut, creeping indigo, jack bean, shortflower rattlebox (crotalaria breviflora), Grant's rattlebox (crotalaria grantiana), rattleweed (crotalaria retusa), cowpeas, etc. Promising winter species are hairy vetch, black oats, ryegrass, corn spurrey, and others. The green manure/cover crops may be in association, or in rotation, with annual crops (Figures 50 and 51).

Species of GMCC that climb, such as mucuna, may also be used as cover crops between the rows of perennial crops, as long as they are pruned periodically to prevent them from climbing and damaging the main crop.

FIGURE 50
The association of cassava with jack bean is one of the alternatives utilized in crop rotations between the rows of yerba mate

FIGURE 51
Bean crop sown No-Till with a manual jab planter between the rows of yerba mate, over soil well covered by black oats that protect it from erosion.

3.6.4 Use of perennial and semi-perennial green manure/cover crops (GMCCs) to recuperate degraded soils

- **In alley-cropping systems with annual crops**

This system consists of planting perennial or semi-perennial green manures (leucaena, pigeon pea, tephrosia) in strips (which may be on the contour), leaving 6 to 12 m between strips (the "alleys") in which the farmer's traditional crops will be grown. The green manures should be cut back periodically (2 to 3 cuts per year) and distributed over the soil surface, to be utilized as cover as well as to add organic matter and nitrogen. Species that are used for this purpose produce great quantities of vegetative mass and have a good ability to resprout. Being shrubby legumes, they develop great quantities of biomass and have deep roots, which carry out important nutrient recycling and nitrogen fixation.

- **On total area as an alternative to fallow**

It is possible to sow species such as pigeon pea, lablab, and tephrosia on degraded soils, leaving them to grow for 2 to 5 years with the objective of recuperating soil fertility. This is proposed as an alternative to the traditional fallow system in which land is not cultivated and left to go to weed for several years. This fallow system favors the multiplication of weeds and infestation by a high quantity of weed seeds, and soil recuperation is generally low due to the scarce quantity of biomass produced.

> The key to success in the use of green manure/cover crops is to find the most appropriate way to include them in production systems on small farms in the different regions.

3.7 ALTERNATIVE USES FOR GREEN MANURE/COVER CROPS

In addition to the protection and improvement of soil, many species of green manure/cover crops may be utilized simultaneously for other purposes.

3.7.1 Animal and human nutrition

Several species of green manure/cover crop serve as forage for animals (cattle, pigs, sheep, chickens, fish, etc.). Depending on the case, they may be grazed directly or utilized as hay, silage, or in rations that include grains.

The species normally used for these purposes are presented in Tables 11 & 12.

GREEN MANURE/COVER CROPS (GMCCS)

TABLE 11
Options for the use of winter green manure/cover crops for animal and human nutrition.

SPECIES	CATTLE		SWINE		POULTRY	HUMANS
	Forage	Grains	Forage	Grains	Grains	Grains
Black oats	Yes	Yes	Yes	No	No	No
Rye	Yes	Yes	Yes	Yes	Yes	Yes
Triticale	Yes	Yes	Yes	Yes	Yes	Yes
Ryegrass	Yes	No	Yes	No	Yes	No
Oilseed radish	Yes	No	No	No	Yes	No
White oats	Yes	Yes	Yes	Yes	Yes	Yes
Forage peas	Yes	Yes	Yes	Yes	Yes	Yes[1]
Hairy vetch	Yes	No	Yes	No	No	No
Common vetch	Yes	No	Yes	No	Yes	No
Sweet lupine	Yes	Yes	Yes	Yes	Yes	Yes
Bitter lupine	Yes	Yes[1]	Yes	Yes[1]	Yes[1]	Yes[1]

[1] Green or dry (toasted) grains.
Source: Adapted from Calegari & Peñalva, 1994.

TABLE 12
Options for the use of summer green manure/cover crops for animal and human nutrition.

SPECIES	CATTLE		SWINE		POULTRY	HUMANS
	Forage	Grains	Forage	Grains	Grains	Grains
Pearl millet	Yes	Yes	Yes	Yes	Yes	Yes
Pigeon pea	Yes	Yes	Yes	Yes	Yes	Yes
Sorghum	Yes	No	Yes	Yes	Yes	No
Forage peanut	Yes	No	Yes	No	No	No
Creeping indigo	Yes	No	Yes	No	No	No
Lablab	Yes	Yes	Yes	Yes	No	Yes
Leucaena[1]	Yes	No	Yes	No	Yes	No
Grey-seeded mucuna	No	Yes	No	Yes[2]	No	No
Dwarf mucuna	No	Yes	No	Yes[2]	No	No
Mung bean	Yes	No	Yes	Yes	No	Yes
Jack bean	No	No	No	Yes[2]	No	No
Sunnhemp	Yes	No	No	No	No	No
Cowpea	Yes	Yes	Yes	Yes	No	Yes

[1] Possesses a toxin (mimosine) that limits the quantity that can be consumed daily.
[2] Possesses toxins, therefore should be treated with heat for its utilization.
Source: Adapted from Calegari & Peñalva, 1994.

Some grains possess toxic substances, and therefore need to be treated with heat or washed with water, etc., in order to be consumed. To be utilized dry they should be toasted in the oven, otherwise they may cause problems for pigs, poultry, and cattle. The common varieties of bitter lupine have high alkaloid content (lupinine, sparteine, lupanine, hydroxylupanine, and anagyrine) and are toxic if consumed by humans or animals. It is possible to eliminate these alkaloids from dry grains by soaking them for 2 days, and changing the water 4 to 5 times to dissolve them since they are concentrated almost exclusively in the skin. There are varieties known as "sweet" with very low alkaloid content in the grains and foliage, which therefore may be destined directly for consumption. However, in Paraguay it is almost exclusively the bitter white lupine that is grown. Grain free of alkaloids may be ground and used as flour, mixed with wheat flour, corn, triticale, etc., to enrich protein content. This product may be used to feed fish, horses, rabbits, etc.

The mucunas should be treated beforehand with heat (cooked for a minimum of 2 hours or toasted) in order to deactivate the L-Dopa, a substance that could be damaging to the organism (World Neighbors, 1989 cited in: Derpsch & Florentín, 1992). There is good experience with rations to feed pigs that are 25% boiled mucuna mixed with corn and cassava.

Jack bean also needs to be cooked for several hours and the water changed (2 to 3 times) to eliminate the effects of toxins (canavaline). The grains of this species, once treated, may be utilized for animal and human consumption.

The sowing of legumes (vetch, lablab, mucuna, etc.) in grass pastures is important to improve the protein content in the animals' diet and to increase forage production. Some legumes may under certain conditions produce tympany or meteorism (accumulation of gases in the belly that can produce death) in ruminants. This problem may be avoided through controlled grazing (few hours), the provision of additional fodder with high fiber content (straw, hay), and/or the utilization of mixtures with grasses.

Mixtures of green manure/cover crops (black oats + vetch, black oats + oilseed radish, pearl millet + cowpeas, and pearl millet + pigeon pea), besides being favorable for the soil and commercial crops, provide a greater volume of forage that is more balanced from a nutritional point of view (energy, protein) than pure crops.

Pearl millet and forage sorghum are outstanding for their high biomass production. They may be utilized in direct grazing or by cutting (2 to 3 times), and the sprouts flattened later as cover (desiccation with herbicides) for the sowing of commercial crops. Forage sorghum has the disadvantage of having a high content of hydrocyanic acid in the early stages of its development, therefore it is recommended that grazing be initiated only at just 35 to 45 days after sowing or sprouting, at which time it no longer causes problems for animals.

Sunnhemp, pearl millet, and pigeon pea may be utilized to recuperate degraded pasture in associated sowings. Besides improving the forage's nutritional value, this makes it possible to improve the soil. Pigeon pea grains serve as food for humans and animals (especially chickens).

Lablab is a biennial, therefore may be counted on for forage over a longer period of time than annual species.

Forage peanut is utilized as forage in countries such as Colombia and Venezuela, principally for cattle but also for sheep. It presents good growth in the summer, with the potential to resprout 2 to 4 times when grazed.

Winter grasses such as white oats, rye, etc., besides being used for animal feed may also be used as flour for human consumption. White oats are used principally as flakes.

In Paraguay, cowpeas are cultivated principally as a source of grain for human consumption and not as a green manure/cover crop. The green seedpods and dry grains are used.

Corn spurrey is good forage, very palatable to animals when it is green or as hay (cattle, swine, sheep, poultry, etc.). It can be cut several times and resprouts well. In dairy animals, the consumption of this forage increases production and gives milk and its products a characteristic flavor that is valued by some consumers.

Leucaena is excellent forage to feed cattle, sheep, poultry, and rabbits. The presence of a toxin (mimosine) in its foliage limits the quantity that can be ingested daily (should not exceed 20% of the diet).

3.7.2 In beekeeping

Some plants show elevated production of nectar and pollen, which can be taken advantage of for beekeeping. The most outstanding species in this aspect are principally oilseed radish and, to a lesser degree, lupines, vetches, sunflower, pigeon pea, lablab, crotalarias, indigo, forage peanut, etc. The presence of these species lengthens the period of time that flowers are available to bees and increases hive production.

3.7.3 Utilization for firewood

Shrubby and arboreal green manures such as pigeon pea, tephrosia, and leucaena may be utilized as firewood.

3.8 RESIDUAL EFFECTS OF GREEN MANURE/COVER CROPS ON MAIN CROPS

The greatest incentive that a farmer finds to use green manures is the immediate residual fertilizer effect on the yield of cash and subsistence crops. However, it is necessary to adequately select the green manures and to insert them in crop rotations in such a way that a greater economic return is achieved.

The advantage of using green manure/cover crops of the legume family is that they have the characteristic of adding great quantities of nitrogen to the soil in a form that is rapidly available, the product of symbiotic fixation. This makes it possible to achieve significant responses in the yield of crops such as corn and cotton (nitrogen demanding) that are sown after legumes. This is principally true of those sown after summer species such as grey-seeded mucuna, jack bean, and pigeon pea, and winter species such as white lupine and hairy vetch, among others.

Oilseed radish, which belongs to the cruciferous family, also brings about a good response in the yield of corn, cotton, and other crops, due to the fact

> By using leguminous green manures, the farmer is installing a nitrogen factory on his/her own farm. Legumes should precede crops that are nitrogen demanding, such as corn and cotton.

that it recycles great quantities of nitrogen that had been washed to deeper layers of soil. Furthermore, it decomposes rapidly after it is flattened, making the nutrients available to crops.

Species of green manure/cover crop that belong to the gramineous family also recycle great quantities of nitrogen and other nutrients, but because of their slow decomposition in comparison with other species, they don't make nitrogen available for the initial development of crops that follow. Furthermore, when great quantities of biomass from a GMCC (gramineous) are left on the soil, microorganisms utilize soil nitrogen to break it down, immobilizing this element. This leads to nutrient deficiencies in the crops, which can reach the point of being severe on soils that are poor in organic matter. As a consequence, there may be an important reduction in crop yields, especially if they are not leguminous. In spite of this problem, gramineous GMCCs, in contrast to those of other families, are excellent for their soil cover that is maintained over a greater period of time, with all the benefits that that implies (reduced erosion, weed suppression, maintenance of soil humidity, etc.).

In order to eliminate a possibly depressive effect on the yield of crops that follow gramineous GMCCs, and to take advantage of the benefits of their cover, several strategies may be used:
- Sow crops of the legume family (cowpeas, peanuts, Phaseolus beans, peas, soybeans, etc.) that have good nitrogen-fixing capacity after gramineous green manures/cover crops.
- Utilize gramineous green manure/cover crops in mixtures with those of other families (principally leguminous and cruciferous). With this practice, when the mixtures are balanced with regard to the population of plants of different species, the effects of cover and nitrogen availability obtained are intermediate between those of pure gramineous GMCCs and those of the other families.
- Open furrows in the rows where the crop will be planted with an animal-drawn ripper (Figures 52-A y 53-A). In this way the cover is pulled back from the proximity of the crop, avoiding nitrogen immobilization. Furthermore, stirring the soil produces greater mineralization (decomposition) of the soil's organic matter, which makes a greater quantity of nitrogen available to the crop. This furrowing (Figures 52-B and 53-B) may at the same time break compacted layers of soil and favor rapid initial development of roots and plants. Cotton sown in No-Tillage responds very favorably to this

practice (Figure 54). In contrast, the lack of cover over the rows favors the proliferation of weeds in the lines, increases the risk of erosion, and increases water evaporation from the rows.
- Increase the lapse of time between when the gramineous green manure/cover crop is flattened and the following crop is sown (leave a period of 4 weeks). This may avoid or even eliminate the effect of nitrogen immobilization. One of the possible disadvantages is that weed infestation could occur, which would require additional weed control.
- Fertilize with nitrogen when sowing the crop (20 to 40 kg/ha of additional nitrogen). This will compensate for the deficiency of this element during microbial immobilization.

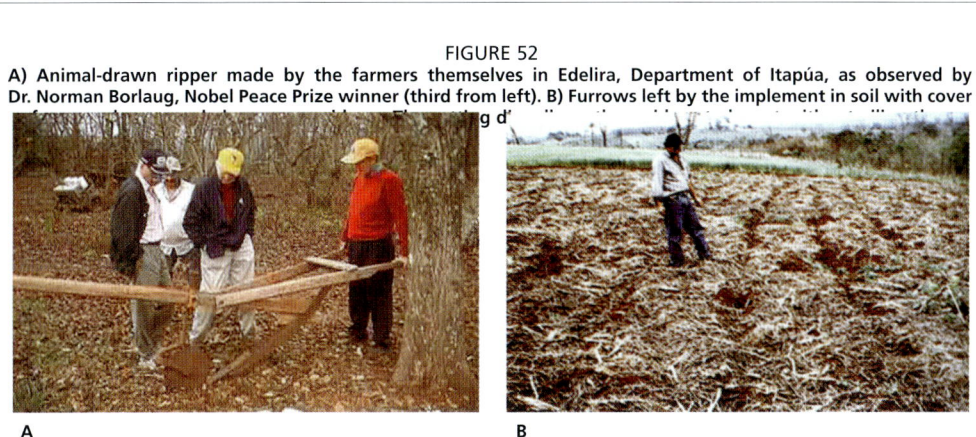

FIGURE 52
A) Animal-drawn ripper made by the farmers themselves in Edelira, Department of Itapúa, as observed by Dr. Norman Borlaug, Nobel Peace Prize winner (third from left). B) Furrows left by the implement in soil with cover

FIGURE 53
A) Animal drawn ripper transformed from a moldboard plow. B) The ripper removes a minimum of cover.

3.8.1 Residual fertilizer effect of green manures on cotton

On sandy soils of medium fertility at the Choré Experimental Station, cotton yields increased with the use of summer green manures associated with corn (Derpsch & Florentín, 1992). The greatest increases in cotton production, in relation to that reached in the conventional system without green manure (after winter fallow), were 55% with the use of grey-seeded mucuna, 52% with black-seeded mucuna, 48% with sunnhemp, and 40% with pigeon pea (Figure 55). The residual effect of winter green manure/cover crops on cotton was much less than that obtained with summer GMCCs (Figure 56).

It is recommended that cotton not be sown on extremely degraded soils, as it is not profitable due to its very low productivity. However, after a soil recuperation plan of at least 2 to 3 years, with the use of green manure/cover crops as an essential component of the system (first pigeon pea and later grey-seeded mucuna, associated with corn), using chemical fertilizers on the crops and utilizing a Conservation Agriculture, it is possible to return to yields greater than 2,000 kg/ha of cotton. This has been proven in the Department of Paraguarí. In this way, cotton may once again be a profitable crop.

FIGURE 54
Cotton sown with a manual jab planter after furrowing (A), shows better development and a yield of 17% more, compared to cotton sown without passing the ripper (B).

FIGURE 55
Residual fertilizer effect of summer green manure/cover crops on the yield of cotton in No-Tillage. Choré Experimental Station. Average of three agricultural years (1989/90, 1990/91 and 1991/92).

Source: Adapted from Derpsch & Florentín, 1992 and Florentín, 1999.

FIGURE 56
Residual effect of winter green manure/cover crops on cotton yields. Choré Experimental Station. Average of three agricultural years (1996/97, 1997/98 and 1998/99).

Source: Florentín, 1999 (unpublished).

3.8.2 Residual fertilizer effect of green manures on corn

In general, corn yields increase significantly when preceded by green manures that add nitrogen, whether through biological fixation (legumes) or by recycling of this element.

On degraded soils in the Department of Paraguarí, on fertilized fields, an increase in corn yields of almost 1,000 kg/ha was obtained, due to the use of a summer GMCC (pigeon pea) that had been associated with corn (Figure 57).

On the Choré Experimental Station, corn production on soils of medium fertility was significantly lower in the conventional system, without GMCC, when compared with production obtained on plots that had had green manure/cover crops associated with corn (Figure 58). Black-seeded mucuna was the summer GMCC that produced the greatest residual effect on corn yield (variety Guaraní V-312) with a production of 5,433 kg/ha, presenting an increase of 56%, in relation to the conventional system that reached 3,479

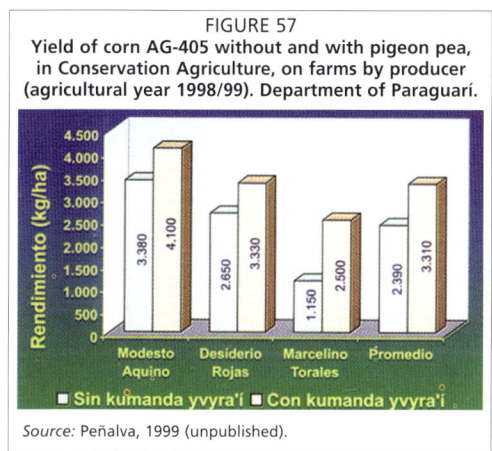

FIGURE 57
Yield of corn AG-405 without and with pigeon pea, in Conservation Agriculture, on farms by producer (agricultural year 1998/99). Department of Paraguarí.

Source: Peñalva, 1999 (unpublished).

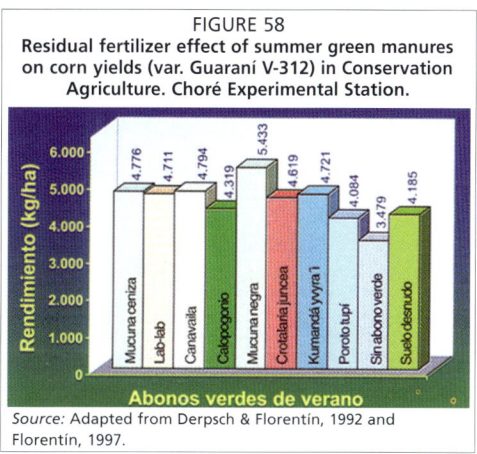

FIGURE 58
Residual fertilizer effect of summer green manures on corn yields (var. Guaraní V-312) in Conservation Agriculture. Choré Experimental Station.

Source: Adapted from Derpsch & Florentín, 1992 and Florentín, 1997.

FIGURE 59
Corn at 24 days after sowing, with little development, sown on a field without a green manure/cover crop after winter fallow (A), compared to well-developed corn over jack bean cover (B).

A

B

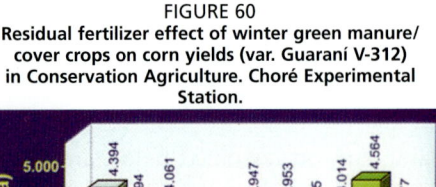

FIGURE 60
Residual fertilizer effect of winter green manure/cover crops on corn yields (var. Guaraní V-312) in Conservation Agriculture. Choré Experimental Station.

Source: Adapted from Florentín, 1999.

FIGURE 61
Tobacco crop, showing good development and without weeds, with good soil cover in the Conservation Agriculture.

kg/ha of corn. The other species studied also showed good residual effects on corn, increasing yields by 38% after jack bean (Figure 59), 36% after grey-seeded mucuna, 35% following pigeon pea, 33% after lablab, 27% after calopo and, finally, by 17% after cowpeas.

On the same experimental station, increases of corn yield were observed after winter green manure/cover crops; the most outstanding was after white lupine, sown alone or in mixture with black oats (Figure 56).

3.8.3. Residual fertilizer effect of green manures on tobacco

Tobacco (Figure 61) is a crop that is very demanding of fertility, principally of nitrogen. On small farms it is traditionally grown on soils recently cleared from forest or, when soil fertility is low, with high doses of chemical fertilizers. On soils of medium fertility, with the exclusive use of green manures for one agricultural season, it is possible to significantly increase the yield of tobacco; however, yields don't reach those normally obtained under the traditional conditions of soil recently cleared from forest.

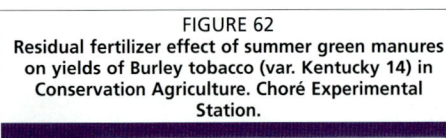

FIGURE 62
Residual fertilizer effect of summer green manures on yields of Burley tobacco (var. Kentucky 14) in Conservation Agriculture. Choré Experimental Station.

Source: Adapted from Derpsch & Florentín, 1992 and Florentín, 1997.

Results obtained on the Choré Experimental Station revealed that the production of Burley tobacco obtained after GMCCs associated with corn was 833 kg/ha in the case of black-seeded mucuna, achieving an increase of 75% in relation to the conventional system without green manure/cover crops (Figure 62). Grey-seeded mucuna and sunnhemp also showed good residual effect on tobacco, resulting in yield increases of 49 and 18%, respectively, in relation to a check plot without GMCC (Figure 63). In crops on land recently cleared from forest, yields normally reach approximately 1,500 to 2,000 kg/ha.

3.8.4. Residual fertilizer effect of green manures on cassava

Due to the fact that its roots are what is harvested, it was doubted that good yields could be obtained by cultivating cassava in No-Tillage. Experiences in the Eastern Region, in sandy as well as clay soils, have demonstrated the viability of cassava cultivation in No-Till systems with green manure/cover crops, such as grey-seeded mucuna and black oats (Figure 64), which leave the soil loose and control weeds, favoring its development (Figure 65). In Edelira, Department of Itapúa, the yield of No-Till cassava over black oats was superior to the average yield in the region (Figure 66).

FIGURE 63
Tobacco planted No-Till over cover of grey-seeded mucuna (foreground, left) with greater development than over the cover of other GMCCs or over winter fallow (background, continuing the rows).

FIGURE 64
Cassava in No-Tillage over cover of black oats, before sprouting (A) and with good initial development (B).

A

B

FIGURE 65
Cassava in No-Tillage over cover of black oats, eight months after sowing and soon before harvest. Good residual effect.

FIGURE 66
Cassava yields (average, 1996-1998) with 12 months of development, grown in Conservation Agriculture over black oats, compared to the average yield for the Department of Itapúa (conventional cultivation without green manure/cover crop).

Source: Adapted from Caballero & Vega1, 1998.

CHAPTER 4
Crop rotation

4.1 GENERAL CONSIDERATIONS

Crop rotation is the alternation of subsistence, cash and green manure/cover crops (GMCCs) with different characteristics, cultivated on the same field during successive years, and following a previously established sequence. The principal objective of crop rotation is to contribute to the achievement of a production that is profitable and sustainable, maintaining soil fertility and health.

In contrast, monoculture – that is to say, the cultivation of the same species year after year in the same place – results in a series of problems that are listed as follows:
- increase of pests and diseases.
- proliferation of certain weeds.
- reduced yields.
- greater economic risk.
- inadequate distribution of labor throughout the year.
- increase in toxic substances or growth inhibitors in the soil.
- reduced biological diversity.

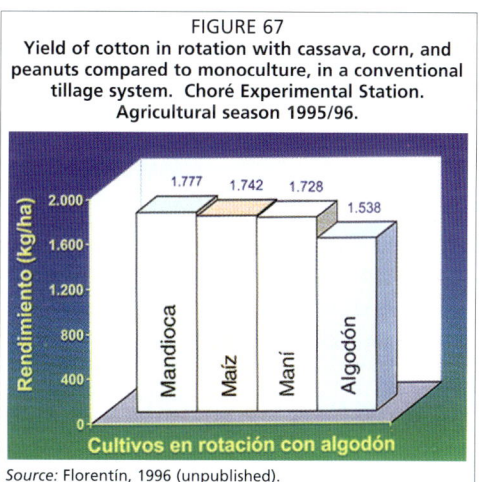

FIGURE 67
Yield of cotton in rotation with cassava, corn, and peanuts compared to monoculture, in a conventional tillage system. Choré Experimental Station. Agricultural season 1995/96.

Source: Florentín, 1996 (unpublished).

In spite of the fact that agricultural systems on small farms are quite diverse, it's common that farmers practice monoculture, principally of cotton. This is because the preference is to utilize this crop on the most fertile soils and the best cleared land, in order to make possible the use of animal-drawn implements.

In an experiment on crop rotation carried out over 9 years on sandy soil at the Choré Experimental Station, Department of San Pedro, it was demonstrated that in a conventional tillage system, without the use of GMCCs, the lowest cotton production was obtained in the treatment with 9 years of monoculture, in relation to plots on which rotations with cassava, peanuts, and corn were utilized (Figure 67).

In the Conservation Agriculture, crop rotation becomes even more important because of the fact that residues of previous crops accumulate on the soil surface. These residues can have a negative effect on crop productivity in the case of monoculture, principally by favoring an increase in pests and diseases (example: cotton), and from possible allelopathic effects from the preceding crop.

In the same experiment shown in Figure 67, but working with green manure/cover crops (GMCCs) and No-Tillage, it was also demonstrated that the lowest yields of cotton, corn, cassava, and peanuts were, in general, obtained on plots where the crop was repeated year after year (Table 13).

For this reason, farmers should abandon monoculture and choose an integrated system of crop rotation and the use of GMCCs in Conservation Agriculture, utilizing corrective fertilization when necessary.

TABLE 13
Production of cotton, cassava, corn and peanuts cultivated for three years in No-Tillage in different rotations, including green manure/cover crops. Choré Experimental Station. Agricultural season 1998/99.

Previous crop	Production obtained (kg/ha)			
	Cotton	Cassava	Corn	Peanuts
Cotton/Black oats+White lupine	740[1]	29.922	4.239	2.422
Cassava/Black oats	1.020	24.388[1]	2.674	2.141
Corn/Grey-seeded mucuna	1.398	28.294	3.507[1]	1.490
Peanuts/Grey-seeded mucuna	1.191	32.385	3.643	913[1]

[1] Monoculture
Source: Florentín, 1999 (unpublished).

There are three fundamental and **basic principles** of crop rotation (Cook and Ellis, 1987):
- Rotation is better than monoculture, even when plants of the same family are cultivated.
- The most efficient rotations are those that include legumes.
- Crop rotation as an isolated practice is generally not enough to maintain stable productivity for many years; the addition of some external nutrients is necessary.

4.2 ADVANTAGES OF CROP ROTATION

Crop rotation presents several advantages compared to monoculture (Mosier and Gustafson, 1917; Cook and Ellis, 1987):
- Better control of insects, other pests and diseases of crops.
- Better weed control.
- Increases crop yields.
- Lower economic risk.

- Better distribution of work throughout the year.
- Reduces the toxicity of several substances in the soil.
- Helps to achieve a more abundant and lasting soil cover.
- Helps to maintain and/or increase soil organic matter content.
- More uniform and stable extraction of nutrients, favoring equilibrium in the soil profile, by alternating root systems with different characteristics (tap, fascicular) and depths.
- Improves soil structure, facilitating crop development.
- Greater biological diversity.

4.3 ASPECTS TO TAKE INTO ACCOUNT IN ORDER TO ESTABLISH CROP ROTATIONS

- Always include green manure/cover crops (GMCCs), prioritizing the production of biomass to improve soil cover and organic matter content.
- The same species should never be sown on the same field in the following season.
- The GMCCs utilized should be adapted to the region's microclimate, to the soil, and to the farmer's production system, and should result in important benefits for cash crops.
- In order to plan crop rotations, the effects of one crop on the following should be taken into account, considering:
  - Compatibility with the following crop.
  - Degree of resistance to attack by pests and diseases.
  - Biomass production.
  - Root system.
  - Nutritional requirements.

High pest and disease infestations have been observed in cotton and peanut No-Till monocultures on the Choré Experimental Station.

> Crop rotation is the most efficient and economical way to break the biological cycles of pests and diseases, thereby making Conservation Agriculture feasible.

With the utilization of green manure/cover crops (GMCCs) in crop rotation, great achievements have been obtained in the maintenance and increase of agricultural production in several Latin American countries, facilitating the diffusion and adoption of Conservation Agriculture on small, medium, and large farms.

4.4 PROPOSED PRODUCTION SYSTEMS WITH CROP ROTATIONS

The proposals were developed utilizing the following aspects as basic fundamentals for the maintenance and recuperation of soil fertility:
- Add to the soil the greatest quantity of organic matter possible.
- Assure permanent vegetative soil cover, live or dead.
- Don't plow the soil; that is to say, practice No-Tillage.
- Use green manure/cover crops.
- Practice crop rotation.
- Apply lime and fertilizers to correct eventual deficiencies and to increase the production of biomass.
- Utilize agrochemicals that are specific or of low toxicity to control pests and diseases, and only when necessary.

4.4.1 Crop rotation in Conservation Agriculture for soils of moderate fertility

Considering that the total agricultural area of the majority of small farms settled over sandy soils of Paraguay's Eastern Region is cultivated with approximately 30% corn and 30% cotton during the summer, it is easy to promote a biennial rotation system with both crops. In addition, the use of GMCCs should be included during periods of time without crops (winter fallow), which can also be achieved by associating GMCCs with traditional crops.

One easy and appropriate option already utilized by farmers is the biennial rotation of corn associated with grey-seeded mucuna, followed by cotton and later winter green manure/cover crops such as bitter white lupine, black oats, and oilseed radish, pure or in mixture (Table 14).

TABLE 14
Two-year crop rotation proposed for moderately fertile soils of the Eastern Region of Paraguay, including green manure/cover crops and No-Tillage.

YEAR	CROPS	
	SUMMER	WINTER
First	Corn / Grey-seeded mucuna	Grey-seeded mucuna continues
Second	Cotton	Black oats + White lupine + Oilseed radish
Third	Corn / Grey-seeded mucuna	Grey-seeded mucuna continues

Grey-seeded mucuna is one of the best summer green manure/cover crop species to be used in association with corn. It is outstanding for its good residual fertilizer effect on cotton, which after mucuna succeeds in producing high yields that are similar to those of cotton with heavy fertilization (Figure 68). Moreover, mucuna offers the advantage of being a species with seed that a farmer can produce on his own land. It is sensitive to frost, and when this occurs the mucuna dies by itself; in this case, there is generally no need to flatten it. This, together with its characteristic as a good weed suppressor, makes it possible through the associated corn/mucuna system to save much labor in the sowing of the following summer crop.

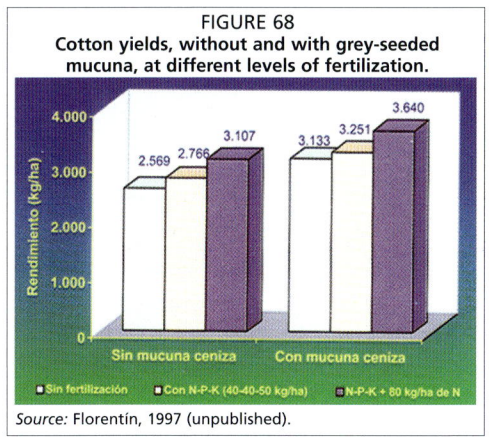

FIGURE 68
Cotton yields, without and with grey-seeded mucuna, at different levels of fertilization.

Source: Florentín, 1997 (unpublished).

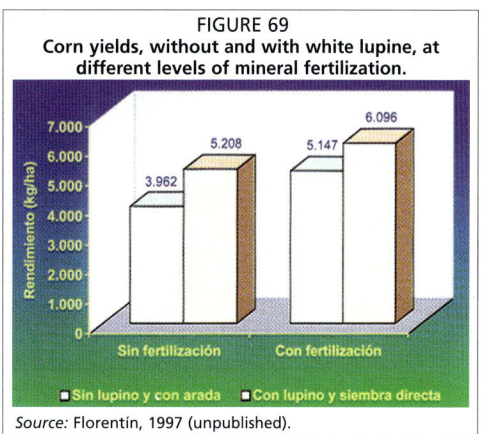

FIGURE 69
Corn yields, without and with white lupine, at different levels of mineral fertilization.

Source: Florentín, 1997 (unpublished).

The species of winter green manure/cover crops recommended to precede corn are a mixture of black oats, white lupine, and oilseed radish, or the same species in pure form, preferably white lupine or oilseed radish. Winter GMCCs, pure or in mixture (associated), bring about significant increases in corn yields, principally when legumes are used (Figure 69).

Winter green manure/cover crops (GMCCs) that follow cotton should be sown, preferably, at the end of March until April, before or after the cotton plant residues are destroyed. The destruction of plant residues should be done with a machete or a knife-roller with additional weight. If necessary, the knife-roller may be complemented by machete, leaving the residues on the field. Cotton plant residues should not be burned, because this impoverishes the soil. Since there is normally a high infestation of weeds at this time, it is necessary to control them with a hoeing or an application of herbicides to allow No-Till sowing of GMCCs. Once established, GMCCs don't need crop management practices or phytosanitary treatment until they are flattened.

The activities to be implemented under a production system of agricultural crops associated with GMCCs, in a corn-cotton rotation, are the following:
- Begin with conventional sowing of corn in September-October.
- Sow 2 rows of grey-seeded mucuna between every two rows of corn (120 kg/ha), 90 to 120 days after the corn was sown, after an additional hoeing.
- Harvest the corn around 40 days after the mucuna was sown.
- If the corn is not harvested, bend the stalks and let the mucuna grow over them until the end of its cycle in July.
- Harvest the mucuna seed to keep it from becoming a weed.

- Pass a knife-roller to crush the corn plant residues and the mucuna.
- If weed infestation should occur, open paths with a hoe in the rows planned for cotton.
- Sow the cotton No-Till with a manual seed drill.
- Continue with the normal cotton production process, taking into account that with this system the first hoeing is avoided.
- After the cotton is harvested, destroy the plant residues by means of a knife-roller with additional weight, complemented with a machete, or with a machete alone if a roller is not available. It's possible to take advantage of the passing of the roller to sow small-seeded winter GMCCs.
- If necessary, eliminate weeds by hoeing or by applying an herbicide (glyphosate, 2-4 D).
- Sow small-seeded GMCCs broadcast and large-seeded ones with a manual seed drill; the following seed quantities are recommended:
  – Black oats + White lupine + Oilseed radish (35+ 60+ 5 kg/ha).
  – Black oats + White lupine (50 + 80 kg/ha).
  – White lupine (80 kg/ha).
  – Oilseed radish (20 kg/ha).
- Secure contact between small seeds and the soil by passing a knife-roller, twice if necessary, or an animal-drawn disc harrow (regulated to roll over the ground, cutting as little as possible into the soil).
- Once the GMCCs are sown, no other activity is required until they are flattened.
- Flatten the GMCCs using a knife-roller, log, or tires, which will end the system's cycle. Starting here, the No-Tillage rotation begins again.
- Optionally, after the winter green manure/cover crop (that was sown after cotton), cassava may be grown in association with jack bean or other summer GMCC, or with winter GMCCs if used for seed. Later, the cycle begins again with corn. Another option is to grow sesame, which may substitute in part or totally the cotton or cassava crops, in association or in sequence with the same green manure/cover crops as cotton (Table 15).

TABLE 15
Three year rotation proposed for moderately fertile soils of the Eastern Region of Paraguay, including green manure/cover crops and No-Tillage.

YEAR	CROPS	
	SUMMER	WINTER
First	Corn / Grey-seeded mucuna	Grey-seeded mucuna continues
Second	Cotton and/or Sesame	Black oats + White lupine + Oilseed radish
Third	Cassava / Jack bean	Cassava continues
Fourth	Corn / Grey-seeded mucuna	Grey-seeded mucuna continues

Economic feasibility of the biennial rotation corn/grey-seeded mucuna – cotton/winter GMCCs in Conservation Agriculture

According to results obtained on the Choré Experimental Station, when cotton production after corn/mucuna was compared to that obtained in the conventional systems used by farmers, there was an increase of approximately 950 kg/ha (average of 3 years of study). Likewise, corn production after white lupine increased by 1,200kg/ha in relation to the conventional system. Based on these results, Duarte and Paniagua (1997) found that the rotation corn/grey-seeded mucuna – cotton/ winter GMCCs had a net income from cotton greater by 836,250 Gs./ha, and from corn production after white lupine greater by 279,200 Gs/ha, than those obtained in the traditional system (Table 16).

TABLE 16
Analysis of system profitability of cotton and corn production in the conventional system (soil tillage, without green manure/cover crops) and in the proposed system (No-Tillage and GMCCs).

Crop rotation	Variable Costs (Gs/ha)	Yield (Kg/ha)	Price (Gs/ha)	Gross income (Gs/ha)	Net income (Gs/ha)
1.Traditional system					
Cotton – conventional tillage	1,320,750	1,736	1,000	1,736,000	415,250
Corn - conventional tillage	705,500	3,962	200	792,400	86,900
2.Proposed system					
First year:					
Corn	705,500	3,962	200	792,400	86,900
Grey-seeded mucuna	360,000	500	1,200	600,000	240,000
Second year:					
Cotton	1,434,500	2,686	1,000	2,686,000	1,251,500
Winter GMCC	207,000				
Third year:					
Corn	575,500	5,208	200	1,041,600	466,100
Grey-seeded mucuna	360,000	500	800	400,000	40,000

Exchange rate (June 1997): 1 dollar = 2,165 guaranies (Gs.)
Data from yields obtained at the Choré Experimental Station.
Source: Adapted from Duarte & Paniagua, 1997.

Apart from the economic benefits that can be obtained on small farms from its implementation, this production system makes it possible to maintain soil fertility, guaranteeing the sustainability of production.

On the other hand, mucuna can generate additional income from the sale of seeds or, if not sold, the farmer can guarantee production of his own seed (opportunity cost). Furthermore, after heat treatment the seed may be used to feed pigs.

Additional equipment required to implement this system is reduced to the acquisition of a knife-roller (Figure 70), a manual seed drill for No-Tillage, and a sprayer to apply herbicides if this doesn't already exist on the property.

Tires (Figure 71) or logs, if available on the small farm, may also be used to flatten some green manure/cover crops (GMCCs).

FIGURE 70
Knife-roller with plates that are interrupted and closer, which softens the impact on the soil and is less tiring for the oxen.

FIGURE 71
An implement to flatten some GMCCs that is economical and easy to make.

4.4.2 Crop rotation in Conservation Agricultures to recuperate extremely degraded soils

In extremely degraded soils there is low recycling of nutrients during the decomposition of organic matter, and low biological activity (small quantity of microorganisms), a situation that occurs principally in the Departments of Paraguarí, Central, Cordillera, and Guairá (Table 17). In these departments, the low productivity of traditional crops (Figure 72), subsistence as well as cash, oblige farmers and their families to dedicate part of their time to other activities such as brick-making, odd jobs on other farms (hoeing, etc.), work in Asunción and other nearby cities, and even outside the country (Buenos Aires). In the last few years, property developers have acquired land in the zones closest to Asunción for subdivision, in large part bought from farmers whose soils produce little.

TABLE 17
Areas cultivated, with extremely degraded soils, and of small farms in the Departments of Paraguarí, Central, Cordillera, and Guairá.

Department	Area cultivated (ha)	Percentage of soils that are extremely degraded (estimated)	Area of extremely degraded soils (ha)	Area of small farms (ha)	Percentage of area of small farms in relation to area cultivated
Paraguarí	96,796	90	87,116	57,643	60
Central	24,217	90	21,795	15,832	65
Cordillera	67,376	90	60,638	41,329	61
Guairá	78,045	90	70,241	45,254	58
TOTAL	266,434	-	239,790	160,058	60

Source: Adapted from Sorrenson, Duarte & López, 2001.

The recuperation of these soils is a gradual process; that is to say, the physical, chemical and biological characteristics of the soil continue to improve in keeping with the application of different management practices (green manure/cover crops, crop rotation, fertilization, No-Tillage, etc.).

In the Department of Paraguarí, where extremely degraded sandy soils exist, a proposal for soil recuperation has been implemented that includes crop rotation with green manure/cover crops in No-Tillage, liming, and crop fertilization (Table 18).

FIGURE 72
Conventional corn crop of very low productivity in the region of extremely degraded soils in the Department of Paraguarí.

TABLE 18
Three year crop rotation proposed for the recuperation of extremely degraded soils in the Eastern Region of Paraguay, including green manure/cover crops and No-Tillage.

YEAR	CROPS	
	SUMMER	WINTER
First	Corn + Pigeon pea	Pigeon pea continues
Second	Corn + Grey-seeded mucuna	Grey-seeded mucuna continues
Third	Cotton or Cassava + Jack bean	White lupine + Black oats + Oilseed radish or Field pea + White lupine + Black oats

When soils are very degraded it is recommended that summer green manure/cover crops (GMCCs) be utilized, as they are more rustic and develop more biomass than winter GMCCs. In this way some of the basic principles of soil recuperation are favored, which are: maximize production of organic matter, improve soil cover, and don't till the soil.

Of the summer GMCCs, pigeon pea was the species that developed the greatest quantity of biomass on extremely degraded soils, due its great rusticity. Furthermore, this species showed itself to be the most feasible for inclusion in production systems on small farms because of the relative ease with which seed is obtained, sown, etc. (Figures 73, 74, and 75). Grey-seeded mucuna did not develop good biomass in the first year of recuperation; even when the corn was fertilized, the corn crop that followed mucuna produced less than with pigeon pea (Figure 76). After soil fertility was raised, with fertilized corn in association with pigeon pea in the first year of recuperation, grey-seeded mucuna showed good development. Therefore, it is recommended that mucuna be associated with fertilized corn as of the second year of soil recuperation.

On extremely degraded soils it is advisable to include winter GMCCs as of the third year (after cotton or associated with cassava), as they are more demanding of fertility than the summer species. The winter GMCCs that perform best are the mixture of black oats, white lupine, and oilseed radish.

Corn is especially suited to initiate the process of recuperating extremely degraded soils, due to its high production of biomass and grain when utilized with green manure/cover crops and chemical fertilizers. Furthermore, this crop leaves a good quantity of plant residues that complement summer GMCCs cover, since its vegetative mass lasts longer. On the other hand, corn makes it possible to sow GMCCs in association, and is traditionally grown on small farms.

FIGURE 73 A AND B
Pigeon pea, sown in association with corn, is recommended to initiate the recuperation of extremely degraded soils, as it is not very demanding of fertility.

FIGURE 74 A AND B
Pigeon pea is cut and flattened around July, when biomass production is high, making total coverage of the soil possible.

FIGURE 75 A AND B
Corn should be sown No-Till over pigeon pea cover 2 to 3 weeks after it was cut. The corn shows good development because it is taking advantage of nutrients released during decomposition of the green manure/cover crop.

A

B

FIGURE 76
Yield of dry matter of pigeon pea and grey-seeded mucuna, and their effect on corn yield (grain and biomass). Sambonini, Department of Paraguarí. Agricultural year 1998/99.

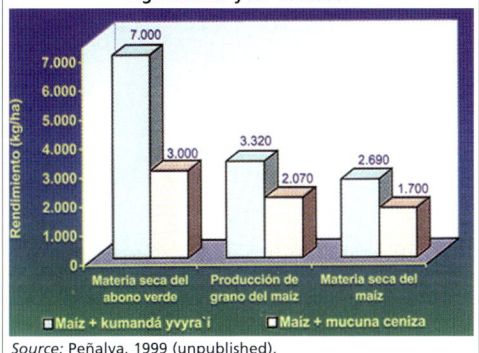

Source: Peñalva, 1999 (unpublished).

FIGURE 77
Corn fertilized with 400 kg/ha of 15/15/15 (N/P/K fertilizer) plus 100 kg/ha of urea on the right, compared with corn that received 200 kg/ha of 15/15/15 (left). On extremely degraded soils, good fertilization of corn is important to reach a high production of biomass and initiate the soil recuperation process.

The utilization of fertilizers on extremely degraded soils is considered important to achieve large quantities of biomass (Figures 77 and 78), not only of the crop being fertilized but also of the green manure/cover crops, which take advantage of residual fertilizer. It also improves crop production, principally of corn which shows a high response to chemical fertilization. This contributes to the improvement of the system's productivity. The fertilizer rates utilized with good results are shown in Table 19. These quantities can be adjusted, based on the needs of each particular farm.

FIGURE 78
On extremely degraded soils, unfertilized gramineous species produce little biomass, as can be observed to the right of the photo. The plant being held on the left was fertilized with 200 kg/ha of 12-12-17-2 (N-P-K-Mg).

FIGURE 79
Yield of hybrid corn (Cargill 805) with two levels of fertilization (average of 6 farms) compared to the conventional system, in the first year of recuperation of extremely degraded sandy soils in the Department of Paraguarí. Agricultural year 1997/98.

Source: Peñalva, 1998 (unpublished).

TABLE 19
Dosages of fertilizers that produced the best response in terms of dry matter, grain, and fiber production on extremely degraded soils in the Department of Paraguarí.

Crop	Nitrogen (N) kg/ha	Phosphorous (P_2O_5) kg/ha	Potassium (K_2O) kg/ha
1st Year: corn[1]	105	60	60
2nd Year: corn	75	30	30
3rd Year: cotton	65	40	50

[1] Limed with 720 kg/ha of limestone PRNT (Relative Total Neutralizing Power) 70%.
Source: Peñalva, 2000 (unpublished).

Figure 79 shows corn's response to fertilization with a compound fertilizer (N-P-K) and urea, in comparison with the yield of conventional corn without fertilization. The increased yield of fertilized corn was highly significant.

One alternative for the recuperation of soils, when financial resources are initially not available to acquire fertilizers, is to grow pigeon pea at high density for 2 to 3 years. After that, it can be left as cover for corn sown No-Till, which can be associated with mucuna or jack bean. In this case, production of pigeon pea seed may provide additional farm income that could be utilized to acquire fertilizer for the corn crop.

Technical and economic feasibility of the proposal for soil recuperation

In the Department of Paraguarí, demonstration plots showed that it is possible to raise the productivity of extremely degraded soils by means of the proposed system. In Yaguarón, Department of Paraguarí, the yield of cotton in the traditional system (with plow and animal-drawn sweep) and with fertilization was only 900 kg/ha. On the same farm a soil recuperation process was initiated on a plot with corn/pigeon pea - corn/grey-seeded mucuna – cotton rotation in No-Tillage, applying lime before the first corn and using fertilizer in the cash crops. In this system cotton production was increased to 2,210 kg/ha (Figure 80).

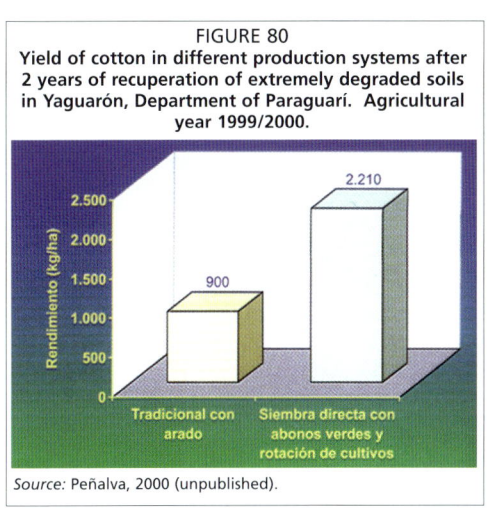

FIGURE 80
Yield of cotton in different production systems after 2 years of recuperation of extremely degraded soils in Yaguarón, Department of Paraguarí. Agricultural year 1999/2000.

Source: Peñalva, 2000 (unpublished).

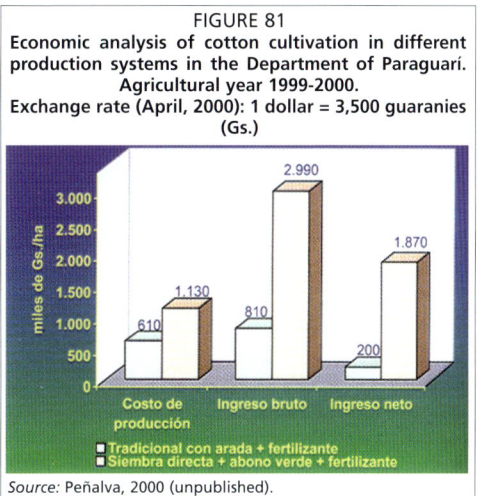

FIGURE 81
Economic analysis of cotton cultivation in different production systems in the Department of Paraguarí. Agricultural year 1999-2000.
Exchange rate (April, 2000): 1 dollar = 3,500 guaranies (Gs.)

Source: Peñalva, 2000 (unpublished).

The economic analysis of cotton cultivation showed a wide advantage of Conservation Agriculture when compared to the conventional (Figure 81). In the system that was implemented, a net income 9 times greater than that of the traditional system was obtained.

In a study carried out by Sorrenson, Duarte & López (2001), on a model farm of 5 ha in the Department of Paraguarí, it was determined that the proposal for the recuperation of extremely degraded soils is a very expensive undertaking for farmers on small properties, therefore incentives (subsidies and credit) are necessary. In the study mentioned, negative net incomes were obtained for the first two years, becoming positive as of the third year (Table 20). It should be clarified that net income under current conditions from the traditional production system is also negative (US$ -176), with the difference that in this situation natural resources are continually degraded and the situation of rural families gradually worsens.

TABLE 20
Economic results from a model farm of 5 ha, before and after financing the recuperation of fertility of extremely degraded soils, with the introduction of Conservation Agriculture in the Department of Paraguarí.

	Total income (US$)	Total costs (US$)	Net farm income (US$)	Total labor (days)
Current	781	957	-176	126
First year[1]	1,015	1,402[2]	-388	167
Second year	1,015	1,159	-145	128
Third year	954	921	33	126
Future	1,621	1,323	298	118

[1] A 75% subsidy is considered for seed, lime and fertilizers in the first year.
[2] This cost includes $150 for the acquisition of 2 manual seed drills and one grain silo.
Exchange rate: (May, 1998): 1 dollar = 2,800 guaranies (Gs.)
Source: Adapted from Sorrenson, Duarte & López, 2001.

The study takes into consideration the purchase of hybrid corn and green manure/cover crop seed. One way to reduce costs for these items is that the farmer sows the corn varieties produced on his own farm, principally Chipá, which normally has a better sale price and produces a good quantity of biomass. On the other hand, the farmer should produce his own green manure/cover crop seed.

4.4.3 Other proposals

• **Sugar cane**

On some small farms on sandy soils of the Eastern Region (Departments of Guairá, Caaguazú, Paraguarí, and Central), sugar cane is the principal cash crop. However, the crop's profitability has fallen, principally because if its low productivity. Low yields are due to the loss of soil fertility from poor management (plowing, burning of residues at harvest, and lack of replacement of nutrients extracted by the harvest).

One of the systems utilized that could be recommended for similar regions is the proposal developed in the Department of Paraguarí, where sugar cane was established in No-Tillage with green manure/cover crops (work conducted by Néstor Paniagua, DEAG-Ybycui).

The system begins with mucuna associated with corn. At the end of mucuna's cycle, furrows are opened at distances of 1.30 m in which sugar cane is planted. Each year after the cane is harvested, 2 rows of dwarf mucuna or jack bean are sown between each two rows of crop.

The proposed system had very positive effects from the production and economic points of view compared to an improved traditional system (soil prepared by plow for plantation, but without burning at harvest). Yields from the first and second cuts were 70 and 60% greater in the proposed system with respect to yields from the traditional system without burning (Figure 82).

CROP ROTATION

In the economic analysis of the systems, it was estimated that net incomes in Conservation Agriculture with mucuna were 7 and 4 times greater in the first and second cuts, respectively, than in the traditional system without burning (Figure 83).

Net income in No-Tillage includes income from the sale of green manure/cover crop seed (grey-seeded mucuna and dwarf mucuna). If GMCC seed is not included, net income in No-Tillage is still greater than in the traditional system without burning.

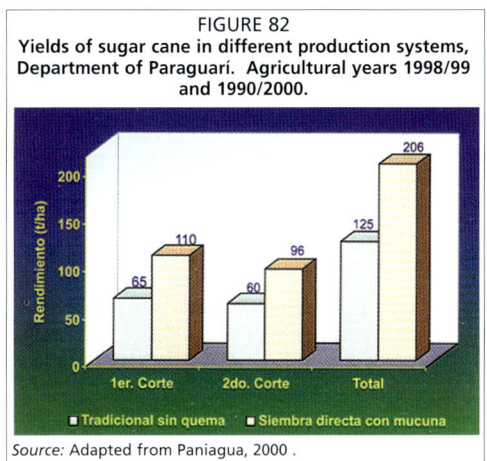

FIGURE 82
Yields of sugar cane in different production systems, Department of Paraguarí. Agricultural years 1998/99 and 1990/2000.

Source: Adapted from Paniagua, 2000.

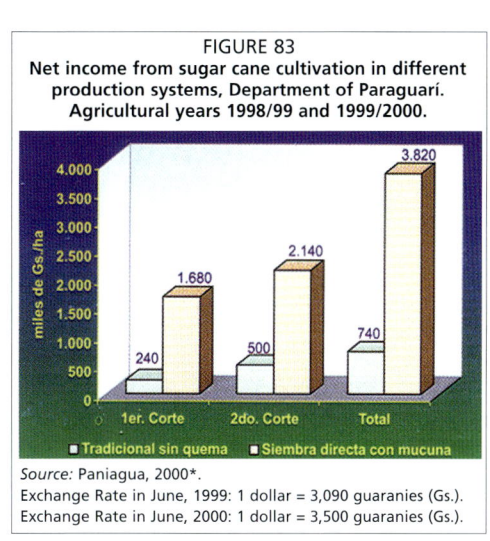

FIGURE 83
Net income from sugar cane cultivation in different production systems, Department of Paraguarí. Agricultural years 1998/99 and 1999/2000.

Source: Paniagua, 2000*.
Exchange Rate in June, 1999: 1 dollar = 3,090 guaranies (Gs.).
Exchange Rate in June, 2000: 1 dollar = 3,500 guaranies (Gs.).

- **Vegetables and horticultural crops (watermelon, melon)**

The use of green manure/cover crops for vegetables and fruits such as melon (Figure 84), watermelon, strawberry, etc., has as its chief objective the formation of dead cover, which makes it possible to maintain soil humidity, reduce temperature extremes, control weeds, and avoid direct contact of fruit with the soil, improving quality. Another important aspect is related to soil health, the association with green manure/cover crops being outstanding for the control of nematodes (crotelarias, mucunas, etc.) and soil fungi (gramineous species such black oats, pearl millet, etc.).

FIGURE 84
Melon in No-Tillage (transplanted into holes) over dead cover of black oats; this improves yields and the quality of fruit, which doesn't rest directly on the soil.

For vegetables that are grown in the fall (lettuce, carrots, etc.), summer green manure/cover crops are the most suitable; good cover is provided by grey-seeded mucuna, pearl millet, sunnhemp, and others, which are sown from October until January. These species, as well as winter GMCCs, pure or in mixture (black oats, hairy vetch, and others), may be left as cover to sow fruits and vegetables that are established as of July-August. When it's necessary to form soil cover in the summer, species that grow rapidly (sunnhemp, pearl millet) may be used.

CHAPTER 5
Advances and obstacles in the adoption of Conservation Agriculture on small farms

A. GREEN MANURE/COVER CROPS

ADVANCES	OBSTACLES
• New species are known and their use has been spread.	• Seeds of some species are difficult to produce.
• The appropriate timing and ways in which green manure/cover crops are placed in production systems are known.	• Difficulties in obtaining seed, due to unavailability or high cost.
• Weeds have been controlled.	• Sometimes seed can't be harvested because labor is assigned to another activity.
• Seed production may serve as a source of income.	• Some seeds are difficult to conserve in good condition due to attack by insects.
• Diversification has increased.	• Training is needed in which of the species are best and in what situations.
• Soils have improved.	
• Yields have increased.	• In some regions there has been little diffusion of the residual effects on crops.
• Pests and diseases have been reduced.	

B. CROP ROTATION

ADVANCES	OBSTACLES
• Pests and diseases have been reduced.	• The most appropriate rotations and their benefits have not been sufficiently disseminated.
• Pesticide costs have been reduced.	
	• In many regions monoculture is prevalent.

C. IMPLEMENTS

ADVANCES	OBSTACLES
• The use of the manual seed drill has facilitated No-Tillage in areas where it hadn't been used.	• Many times seed is not covered well when sown with a manual seed drill; additional work is needed to cover the seed.
• The availability of animal-drawn seed drills that allow for more rapid sowing (making it feasible to work on larger areas).	• Some implements imply a high investment.
	• In some regions the implements needed are not available.

D. WEED CONTROL

ADVANCES	OBSTACLES
• Control has improved. • The need for labor is reduced by the reduced need for hoeing or herbicide use. • Costs have been reduced.	• Knowledge is lacking about integrated weed management. • Training is needed on the use of herbicides. • Herbicides are costly (sometimes there are no economic or financial resources available). • Herbicides are more expensive if bought in small quantities. • Herbicides have restricted availability in some regions.

E. NO-TILLAGE

ADVANCES	OBSTACLES
• Dissemination in pilot zones in all of the Eastern Region. • Soil conservation has improved. • Economic advantages have been demonstrated. • Farmers have more time for other activities. • Some lines of credit have been created for the promotion of No-Tillage.	• Official entities often promote tillage (plowing, disking), especially for the cultivation of cotton. • Lines of credit exist for the purchase of plows and for tillage.

F. CROP PRODUCTIVITY

ADVANCES	OBSTACLES
• Yields have increased and stabilized. • The potential exists for greater yield increases due to gradual improvement in soil fertility.	• In some cases yields have gone down due to lack of information about system management (for example, difficulty in weed control, inadequate selection of green manure/cover crops).

G. FERTILIZERS AND PESTICIDES

ADVANCES	OBSTACLES
• The use of fertilizers on extremely degraded soils accelerates adoption of Conservation Agriculture. • Something has been learned about biological pest control, reducing the use of pesticides.	• Pesticides are costly (sometimes there are no economic or financial resources available for their purchase) and they are more expensive if bought in small quantities. • In many regions fertilizers are not available. • Little dissemination and lack of knowledge about biological pest control.

H. CREDIT

ADVANCES	OBSTACLES
• In some regions, special lines of credit have been obtained for implementation of Conservation Agriculture with green manure/cover crops.	• It is difficult to obtain adequate lines of credit (limited availability, untimely, short-term financing). • Credit is available in few areas for the purchase of inputs (seed drill, knife-roller, etc.).

I. TRAINING

ADVANCES	OBSTACLES
• Demonstration plots (Figure 85) and model farms make it possible to organize field trips that are motivating and generate ideas to apply later on the farm. • More farmers and technicians (multiplying agents) with more knowledge about Conservation Agriculture (Figure 86).	• Within the universe of 260,000 producers on small farms in Paraguay, very few farmers are trained in, and/or are practicing, Conservation Agriculture. • There are not yet enough trained technicians.

J. CULTURE

ADVANCES	OBSTACLES
• Participation of the family, especially of women (Figure 87). • Well-established groups advance more quickly.	• The use of the plow and burning are deep-rooted traditions. • There exists little culture of conservation of natural resources. • It is difficult to change the mentality, especially of those that are older (resistance to change).

K. INVESTIGATION

ADVANCES	OBSTACLES
• Knowledge has been generated for different environments. • Have worked on farms and with farmers. • Farmers themselves have generated important knowledge.	• Investigation is lacking on No-Tillage for intensive crops (horticulture). • Few resources are dedicated to soil conservation and Conservation Agriculture. • Multidisciplinary focus is lacking.

L. SYSTEM PROFITABILITY

ADVANCES	OBSTACLES
• The profitability of the system has been demonstrated. • It is possible to achieve short- and medium-term profitability.	• On extremely degraded soils the initial investment (inputs) is high and generates negative incomes (subsidies and/or special credits are needed).

M. ENVIRONMENT

ADVANCES	OBSTACLES
• It has favored an increase in fauna, in general. • The utilization of pesticides has been reduced, therefore so has the risk of contamination. • Products of lower toxicity and greater specificity are utilized. • Increases the conservation of soil and water.	• Lack of knowledge about the correct use of herbicides and adequate management of containers may lead to contamination of the environment.

N. FARM SIZE AND TENANCY

ADVANCES	OBSTACLES
• Alternatives have been generated to implement the system on small farms (crops associated with green manure/cover crops).	• A farmer does not want to improve the soil when the farm does not belong to him.

O. FAMILY AND SOCIETY

ADVANCES	OBSTACLES
• There is more time to dedicate to other activities. • Reduces rural migration toward the city. • Allows schooling to be improved (greater attendance in schools and other centers of study). • Improves the quality of life.	• Social pressure against innovators (intentional burning of crop residues, exclusion).

P. SUSTAINABILITY

ADVANCES	OBSTACLES

• There are indicators that show that sustainable agriculture may be achieved with Conservation Agriculture:
  - Improves soil fertility.
  - Increases levels of soil organic matter.
  - Increases yields.
  - Increases profitability.
	• None known.

FIGURES 85 A AND B
Demonstration plots conducted by the farmers themselves, with the support of technicians, allow for more rapid dissemination of technologies.

A

B

FIGURE 86
Ongoing training of farmers is essential for the adoption of soil cover systems.

FIGURE 87
Family participation is fundamental for the adoption of Conservation Agriculture on small farms.

Annex

SPECIES OF CULTIVATED PLANTS AND WEEDS MENTIONED IN THIS PUBLICATION:

Common Name	Scientific Name
Bean	*Phaseolus vulgaris* L.
Butterfly Pea	*Clitoria ternatea* L.
Calopo	*Calopogonium mucunoides* Desv.
Cassava	*Manihot esculenta* Crantz sin. *Manihot utilissima* Pohl
Corn	*Zea mays* L.
Corn Spurrey	*Spergula arvensis* L.
Cotton	*Gossypium hirsutum* L.
Cotton Grass	*Digitaria insularis* (L.) Mea ex Ekman
Cowpea	*Vigna ungüiculata* (L.) Walp. sin. *Vigna sinensis* (L.) Savi
Cowpea var. Tupí	*Vigna ungüiculata* (L.) Walp. sin. *Vigna sinensis* (L.) Savi var. Tup'í
Cowpea, Red	*Vigna ungüiculata* (L.) Walp. sin. *Vigna sinensis* (L.) Savi var. Colorado
Creeping Indigo	*Indigofera endecaphylla* L.
Dayflower	*Commelina benghalensis* L.
Forage Sorghum	*Sorghum sp.*
Hispid Starburr	*Acanthospermum hispidum* DC.
Jack Bean	*Canavalia ensiformis* L. DC
Jamaican Crabgrass	*Digitaria horizontalis* Willd.
Lablab	*Lablab purpureum* L. Sweet
Leucaena	*Leucaena leucocephala* L. de Wit
Lupine, Bitter	*Lupinus spp.*, variedades amargas
Lupine, Bitter white	*Lupinus albus* L., var. Amarga
Lupine, Sweet	*Lupinus spp.*, variedades dulces
Lupine, White	*Lupinus albus* L.
Melon	*Cucumis melo* L.
Mucuna	*Mucuna spp.*
Mucuna, Black-seeded	*Mucuna aterrima* = *Stizolobium aterrimum* Piper et Tracy
Mucuna, Dwarf	*Mucuna pruriens* = *Stizolobium deeringianum* Bort.
Mucuna, Grey-seeded	*Mucuna pruriens* = *Stizolobium cinereum*
Mung Bean	*Vigna radiata* L.
Nutgrass	*Cyperus rotundus* L.
Oats	*Avena spp.*
Oats, Black	*Avena strigosa* Schreb
Oats, Black IAPAR 61	*Avena strigosa* Schreb var. IAPAR 61
Oats, White	*Avena sativa* L.
Oats, Yellow	*Avena byzantina* C. Kock
Painted Spurge	*Euphorbia heterophylla* L.

Common Name	Scientific Name
Pea	*Pisum sativum* L.
Pea, Field	*Pisum sativum subesp. Arvense* L.
Peanut	*Arachis hipogea* L.
Peanut, Forage	*Arachis pintoi* L.
Pearl Millet	*Pennisetum americanum* L.
Pigeon Pea	*Cajanus cajan* L. Millsp.
Radish, Oilseed	*Raphanus sativus* L. var. oleiferus Metzg.
Radish, Wild	*Raphanus raphanistrum* L.
Rattlebox	*Crotalaria paulina* Schrank
Rattlebox, Grant's	*Crotalaria grantiana* Harv.
Rattlebox, Shortflower	*Crotalaria breviflora* DC (*Crotalaria divergens* Bth.)
Rattlepod, Streaked	*Crotalaria striata* DC
Rattleweed	*Crotalaria retusa* L.
Rye	*Secale cereale* L.
Ryegrass	*Lollium multiflorum* Lam
Southern Sandbur	*Cenchrus echinatus* L.
Soybean	*Glicine max*
Sugar Cane	*Saccharum officinarum* L.
Sunflower	*Helianthus annuus* L.
Sunnhemp	*Crotalaria juncea* L.
Tephrosia	*Tephrosia tunicata* L., *Tephrosia candida* L.
Tobacco	*Nicotiana tabacum* L.
Triticale	*X Triticosecale* Wittmack
Tropical Mexican Clover	*Richardia brasiliensis* Gomes
Vetch	*Vicia spp.*
Vetch, Common	*Vicia sativa* L.
Vetch, Hairy	*Vicia villosa* Roth
Watermelon	*Citrullus vulgaris* Schrad
Wheat	*Triticum aestivum*
Yerba Mate	*Ilex paraguariensis*

References

AMADO, T. J. C., 2000: Manejo da palha, dinâmica da matéria orgânica e ciclagem de nutrientes em plantio direto. In: 7º Encontro Nacional de Plantio Direto na Palha, Foz do Iguazú, Brasil, Resumos. pp. 105-111.

BUNCH, R., 1995: Principles of agriculture for the humid tropics : An odyssey of discovery. ILEA. Newsletter, October, 1995. pp. 18-19.

CALEGARI, A., 2000: Rotação de culturas e uso de plantas de cobertura : Dificuldades para sua adoção. In: 7º Encontro Nacional de Plantio Direto na Palha, Foz do Iguazú, Brasil, Resumos. pp. 145-152.

CALEGARI, A., 1999: Uso de abonos verdes en Siembra Directa en pequeñas propiedades. In: VIEDMA, L., Coord. Curso de Siembra Directa en Pequeñas Propiedades. MAG/DIA, PROCISUR, GTZ. Encarnación, Paraguay. pp. 29-43.

CALEGARI, A., 1997: Importancia de la rotación de cultivos y abonos verdes en la siembra directa. In: VIEDMA, L., Coord. Curso sobre Siembra Directa. MAG/DIA, PROCISUR/BID. Encarnación, Paraguay. Centro Gráfico. p. 51-68.

CALEGARI, A., 1997: Eficiencia del sistema de Siembra Directa a través del uso de abonos verdes y rotación de cultivos. In: 5º Congreso Nacional de AAPRESID, Mar del Plata, Argentina, Conferencias. pp. 133-151.

CALEGARI, A., 1990: Plantas para adubação verde de inverno no Sudoeste do Paraná, IAPAR, Londrina, Boletim técnico 35. 97 pp.

CALEGARI, A.; FERRO, M.; GRZESIUK, F. and JACINTO JUNIOR, L.,: Plantio direto e rotação de culturas. Experiência em latossolo roxo, 1985-1992. Cooperativa de Cafeicultores e Agropecuaristas de Maringá LTDA., Maringá, 64 pp.

CALEGARI, A.; MONDARDO, A.; BULISANI, E. A.; WILDNER, L. DO P.; COSTA, M. B. B.; ALCÂNTARA, P. B.; MIYASAKA, S. and AMADO, T. J. C., 1993: Adubação verde no Sul do Brasil. Coordenação: M. Baltasar B. da Costa. Rio de Janeiro, AS-PTA, 2. ed. 346 pp.

CALEGARI, A. and PEÑALVA, M., 1994: Abonos verdes : Importancia agroecológica y especies con potencial de uso en el Uruguay. Canelones, MGAP/JUNAGRA–GTZ, 172 pp.

CARDOSO, R. M. L. and TEDARDI, C. R., 1990: Effect of seed processing in potential reduction of *Colletotrichum Gloeosporoides* inoculant in *Lupinus albus* cv. "Vega". In: 6th International Lupin Conference. Proceedings, 1990. Temuco, Pucon, Chile. ILA, pp. 25-30

COOK, R.L. and ELLIS, B.G., 1987: Soil management: A world view of conservation and production. Jhon Wiley and Sons, USA

DERPSCH, R., 1998: Importancia de la rotación de cultivos en el sistema de Siembra Directa. In: 3° Encuentro Nacional de Productores en Siembra Directa, Obligado, Paraguay, pp. 71-102.

DERPSCH, R., 1994: Estrategias de rotaciones de cultivos en el sistema de Siembra Directa: Fundamentos. In: 3° Congreso Nacional de AAPRESID, Villa Giardino, Argentina, Trabajos presentados. pp. 214-247.

DERPSCH, R. and CALEGARI, A., 1985: Guia de plantas para adubação verde de inverno. IAPAR, Londrina, Documentos IAPAR, 9. 96 pp.

DERPSCH, R. and CALEGARI, A., 1992: Plantas de adubação verde de inverno no Paraná. 2a Ed. IAPAR, Londrina, Circular, 73. 80 pp.

DERPSCH, R. and FLORENTÍN, M., 1992: La mucuna y otras plantas de abono verde para pequeñas propiedades. Asunción, Paraguay, MAG, Miscelánea, No. 22. 44 pp.

DERPSCH, R; FLORENTÍN, M. and MORIYA, K., 2000: Importancia de la Siembra Directa para alcanzar la sustentabilidad agrícola. "Proyecto Conservación de Suelos MAG-GTZ". San Lorenzo, Paraguay, 40 pp.

DERPSCH, R., ROTH, C. H., SIDIRAS, N. and KÖPKE, U., 1991: Controle da erosão no Paraná, Brasil: Sistemas de cobertura do solo, plantio direto e preparo conservacionista do solo. GTZ, Eschborn, S. P. 245. 274 pp.

DIRECCIÓN DE CENSO Y ESTADÍSTICA, 1991: Censo Agropecuario Nacional. Dirección de Censo y Estadística, Asunción,. V. 3. 269 pp.

FLORENTÍN, M., 1999: Uso de abonos verdes en los sistemas de producción de los pequeños productores de San Pedro, Paraguay. In: VIEDMA, L., Coord. Curso de Siembra Directa. MAG/DIA, PROCISUR, GTZ. Encarnación, Paraguay. pp. 44-57.

FLORENTÍN, M., 1997: Efecto residual de los abonos verdes sobre la infestación de malezas y la producción de cultivos comerciales. In: VIEDMA, L., Coord. Curso sobre Siembra Directa. MAG/DIA, PROCISUR/BID. Encarnación, Paraguay. Centro Gráfico. pp. 69-84

GASSEN, D. N. and GASSEN, F. R., 1996: Plantio direto, o caminho do futuro. Passo Fundo, Aldeia Sul, 207 pp.

IAPAR., 1998: Plantio direto, pequena propriedade sustentável. IAPAR, Instituto Agronômico do Paraná, Londrina, Brasil, Circular, n. 101. 255 pp.

MONEGAT, C., 1991: Plantas de cobertura do solo : características e manejo em pequenas propriedades. Chapecó, Brasil. Ed. do Autor, 337 pp.

MOSIER, J. G. and GUSTAFSON, A. F., 1917: Soil physics and management. Philadelphia and London, J. B. Lippincott Company, Illinois, USA,

PEÑALVA, M. and CALEGARI, A., 1999: Abonos verdes como integrantes de sistemas de producción hortícolas y frutícolas. Canelones, Uruguay, MGAP/JUNAGRA–GTZ, 154 pp.

SALTON, J. C., 1998: Sistema plantio direto: O productor pergunta, a EMBRAPA responde. Brasilia, Empresa Brasilera de Pesquisa Agropecuária. Serviço de Produção de Informação, Coleção 500 Preguntas 500 Respostas. 248 pp.

SALTON, J. C.; PITOL, C.; SIEDE, P. K.; HERNANI, L. C. and ENDRES, V. C., 1995: Nabo forrageiro : sistemas de manejo. EMBRAPÀ, Dourados, Centro de Pesquisa Agropecuária do Oeste, EMBRAPA –CPAO. Documentos, n. 7. 23 pp.

SEAA, 1994: Manual de uso, manejo e conservação do solo e da água : Projeto de recuperação, conservação e manejo dos recursos naturais em microbacias hidrográficas. Secretaria de Estado da Agricultura e Abastecimento, SEAA, Santa Cararina, Brasil. 2a Ed. Florianópolis, Brasil, EPAGRI, pp. 189-202.

SANTOS, H. P. dos; REIS, E. M. and DERPSCH, R., 1993: Rotação de culturas. In: Plantio Direto no Brasil, CNPT-EMBRAPA, FUNDACEP-FECOTRIGO, FUNDAÇAO ABC. Editora Aldeia Norte, Passo Fundo, pp. 85-103.

SER., 1996: Estudio base de identificación y caracterización de la población meta del proyecto "Conservación de Suelos" MAG-GTZ. I y II. Sociedad de Estudios Rurales y Cultura Popular, SER, Asunción, Paraguay, 98 pp.

VILLALBA, M. A.; LÓPEZ, O. and BOGADO, E. L., 2000: Degradación de los suelos agrícolas de la Colonia Bertoni de San Estanislao. Asunción, Tesis de grado-Carrera de Ciencias Ambientales, Universidad Técnica de Comercialización y Desarrollo.

SORRENSON, W. J.; DUARTE, C.; and LÓPEZ, J., 2001: Aspectos económicos de los sistemas de Siembra Directa y labranza convencional en pequeñas fincas del Paraguay: Implicancias en la política y la inversión. Traducción: H. Causarano, "Proyecto Conservación de Suelos MAG-GTZ", San Lorenzo, Paraguay, 88 pp.